Bootcoder's Complete Guide to Programming Bootcamps

by Henry M. Lewis

Table of Contents

"I have no understanding as to why, when you're 18 to even, let's say 30, why you wouldn't try to make what you're passionate about work for you... If you go and become a lawyer or go to school and do all the things that everybody wants you to do, and don't do the thing you really love, the real question isn't what's going to happen when you're 23, 27, 31, 36. The question really becomes what's going to happen when you're 70 years old and you look back at your life and you're like, 'Why didn't I try?' There's going to be a regret factor that I think a lot of times a guidance counselor or parent or teacher tend not to think about... They're worried about your next ten years. I'm worried about your last ten years. And in those last ten years, you're going to be thinking back... and realizing, 'Why didn't I go to Austin (or L.A. or Nashville... wherever you're going)? Why didn't I take a chance?' and really regret that. And that - that tastes a lot worse than going for it, because that's when you're most alive."

- Gary Vaynerchuk

Preface

People often find themselves at crossroads in their lives. That's why there are such commonly used metaphors as:

> "Here I am at these crossroads again...",
>
> "At every crossroads on the path that leads to the future...",
>
> "I was in the middle of a crossroads...",
>
> The list of memorable quotes is infinitely long. So there

it is - you've found yourself at a metaphorical crossroad, on the wrong street with lots of streets running parallel, in some maze-like labyrinth, or maybe even at a dead end, and you're asking yourself "What's next?"

People place their value in things that often don't or should not define them, and whether that's intellect, finances, relationships, or vocation, thinking that things will ultimately fill some existential hole is unrealistic and will leave said person feeling disappointed. Thus, this book does not claim that becoming a programmer is the be-all, end-all.

Instead, this book points out a unique place in history, an occurrence that happens for all generations, but in different ways. It could be compared to the oil rigs of North Dakota where they are currently fracking away, and in the 1980s it was accountancy, and during the early 20th century it might have been the workers in Ford's factories. What happens is that, regardless of the general state of the economy, some particular industry receives such incredible demand and experiences such explosive growth and makes so much money that it can't keep up with demand, and will pay significantly higher than average wages to workers possessing the necessary skills to perform crucial (although not necessarily complex) roles in the industry. Often, such industries will train workers possessing only the bare minimum of these sought-after skills.

Meritocracy: a system in which the talented are chosen and moved ahead on the basis of their achievement; a system where advancement is based on intellectual talent measured through examination and/or demonstrated achievement in the field where it is implemented; any group of people whose progress is based on ability and talent rather than on class privilege or wealth.

Today, tech has become *the* meritocracy, where *getting in* has become as easy as knowing the fundamentals of program... the rest can be taught. Companies everywhere are experiencing the "tech crunch", and "there is a war in the [tech industry]– a war for talent." This "war" is driving up the perks of being a programmer, which can now include massive signing bonuses, salaries, cool work environments or "cultures", free meals, and months of vacation time. It has truly become the Golden Age of the Programmer.

The Whitehouse recently stated that, "While a college education remains a worthwhile investment overall, the average borrower now graduates with over $26,000 in debt. Only 58 percent of full-time students who began college in 2004 earned a four-year degree within six years. Loan default rates are rising, and too many young adults are burdened with debt as they seek to start a family, buy a home, launch a business, or save for retirement". At the same time, "at too many schools, tuition is going up, graduation rates are going down, and students are leaving with enormous debt and little hope of high-paying jobs." Everyone knows people with college degrees, often masters and PHD level degrees, who are working menial jobs, while shouldering crippling debts. Something needs to be done, and rapid change has been occurring all the way down to the roots of education.

While all of the experts interviewed in this book do not consider programming bootcamps a replacement for a college education, the value of which is intangible, "the college degree has come under a lot of attack as the student debt figures are rising and the percentage of unemployed graduates are being shared with the public...the bootcamp is not a replacement for the college education. Attending a programming bootcamp is a discreet decision...our purpose...is to help get somebody a job. That is a very specific goal, which is not being met by colleges. The decision to attend college is going to be informed by a lot more factors than just getting a job."

If getting a job is your goal, then programming might be right for you. This book will guide you in your expectations for attending a programming bootcamp, and provide invaluable feedback on pretty much everything you need to know to make an informed decision.

Perhaps all of this vocational searching boils down to happiness. You've found yourself asking, "Am I happy?" and realized that the answer might not be 100% "yes". In fact, it is likely that your happiness percentage is much less than that, and that you attribute that largely to whatever career path you are currently on. I am not going to propagate some whimsical concept that "money doesn't buy happiness", because recent studies, and one in particular from *The Economist*, indicate that "the relationship between income and well-being" matters up to a point, and that "people everywhere report more satisfaction as they grow rich", up to around $70,000 USD per year, at which point money usually only gives a person increased comfort, but not necessarily happiness.

What I mean to say is that money cannot buy happiness after a point, but that, like in Maslow's hierarchy of needs, a person will feel increasing "self-actualization", or some sort of greater happiness, as his or her needs are met, allowing that person to move into greater realms of creativity, love, etc. Being bogged down at a crappy job where you don't use your brain, are struggling to pay rent, and don't have any extra cash for eating out or going to the opera or museum or some cultural event sucks. I know. So part of the philosophy of this book *is* that getting a job where you get to think a lot more and be creative and make a decent living will ultimately lead to a better lifestyle on the whole, but it's still going to be different for each individual.

One thing that this book certainly cannot do for you is *CODE*. If you're seriously considering a programming bootcamp and you haven't touched a line of code, you need to go to one of the numerous online tutorials (see the programming resources page) and spend a good ten hours minimum, preferably over a few days, learning Ruby or Python or Javascript or whatever, and then come back here for your next steps in the journey.

Why Software is Eating the World

All around us, software is eating the world. You see it in the street lights you wait at, the restaurants you frequent, the shoes you wear. The goal of the tech industry for a few decades now has been to shift everything up to the internet. Ancient corporate beasts, like the insurance and shipping industries, have been transformed (in part) from storage houses for paperwork with backed up information to finely tuned data-warehouses. The software of early adopters, which was created before you knew what a computer was (or were born), is being recycled into newer, better software. The market is adapting rapidly and lines of code never shipped as quickly as today. The mentality has changed to "If you run a business, put it on the internet". Marc Andreesen, who created Netscape, among other accomplishments, points out that "More and more major businesses and industries are being run on software and delivered as online services—from movies to agriculture to national defense. Many of the winners are Silicon Valley-style entrepreneurial technology companies that are invading and overturning established industry structures." You don't have to be in the loop to see it; what Andreesen points out is something that everyone has observed and experienced in his or her own lifestyle's change over the last ten years. Don't believe me? Think about this: "The Chevy Volt...has 10 million lines of code, 2 million more than the code in the F-35 fighter jet". This represents the transfer of technology to the masses. Also, when was the last time you used a map? Or a phone that had a cord? Or didn't use your email for two days?

Different historical events have made people skeptical of technology's ability to solve all of our problems and meet every need. For example, stock market glitches, such as the initial error on Facebook's IPO or the "Flash Crash of 2010", where an error in an algorithm used by a major firm dropped the share price of the DOW by 1,010.14 points in an instant, have caused skepticism towards technology. Furthermore, pop culture has pretty much accepted that at some point robots will destroy humanity, and everyone is aware that they are spending far too much time on social media and their phones, to the point that posture has become increasingly bad and "iPhone Neck" is a commonly used term.

For all of those reasons, there is a great deal of debate over whether technology is making our lives better. Some think that it is just making us lazier. Probably true. But who wouldn't want to make a robot do the hard work for them, adding up thousands of brain-numbingly easy math equations, or copying and pasting text from one file to another? Unless you're some sort of masochist, technology can and already has made your life easier.

For that very reason "we are in the middle of a dramatic and broad technological and economic shift in which software companies are poised to take over large swathes of the economy." The potential is huge, and the industry to be in right now is tech. There's no way around it. As I write, Internet behemoth Alibaba, which is the world's largest online commerce company, is going public. Right now its share price is 90.79, with a market cap of $230 billion. One of Alibaba's main services is connecting the export industry, typically an outdated and bureaucratic business, with companies from all over the world. It represents yet another industry being transformed by technology.

The entrepreneurial community has noticed too. Since tech became a thing, one of the largest setbacks to starting a tech company has been cost. An early entrant into the technology market comments that tech was, at the time, "kind of a niche" and that "software engineers had to work in big companies who could invest in mainframe computers." Things couldn't be more different now. TechCrunch notes that, "after 2001...the cost of bandwidth, processing power and storage dropped dramatically, and reliable open source computing stacks began to emerge, making it increasingly cheap and easy to launch web services. As broadband penetration increased, the Internet became a much more powerful communication and distribution network. With the advent of paid search and social networking, it became easier than ever for web services companies to find new users and customers. What followed was a period of explosive growth for a new generation of web services — both consumer and enterprise — that took advantage of these new technologies." The ease of entrepreneurial creation has become absurd. If you have an idea, you can hack on it and ship it in a weekend (a.k.a. "hackathon"). Also, the costs are continually decreasing from their already dirt-cheap levels. It is not uncommon for techies today to bootstrap their own business, because it can be done with even a moderate salary.

One Silicon Valley entrepreneur, Christian Gheorghe, runs his 62 million dollar company on cheap and free services, taking advantage of free email, various open source database and word-processing systems (free), and extremely cheap cloud computing with Amazon Web Services (12 cents an hour for the whole company). Furthermore, cloud computing prices keep dropping "every ten weeks or so", (Amazon claiming it has had 17 price drops in the past four years), as are other prices, such as the cost of marketing and market research. A recently launched startup spent about $300 in online advertising in order to understand what 50,000 people in its target market were most interested in. Instead of hiring an expensive PR firm, companies now focus on social media sites to get the word out at a much lower cost. Even logos and other branding can be designed at a fraction of the cost, since websites have emerged that allow designers to submit their business so that the buyer can choose from an array of designs.

As a result, massive tech hubs are starting in cities, people are moving in droves to strike it rich like they were during the gold rush, and new things are being created. The Startup Genome Report is in agreement when it states that the "last ten years has seen a near-total collapse of the innovation cost curve, thanks to the perfect storm of open-source, cloud infrastructure, and "free" global distribution via search, social and app stores. At the same time we have seen the start-up ecosystems in New York, London and Berlin emerge as meaningful competitors to Silicon Valley." It has become almost a cliché for cities to say that they are trying to propel their tech community forward, because it can be such a stimulus for local economies.

Furthermore, technology is often paired with entrepreneurship. Running "entrepreneurship and technology" through your favorite search engine brings up various universities' websites, which are offering courses such as "Technology Entrepreneurship", "Fundamentals of Technology Start-Up Ventures" and "Strategy for High-Tech Ventures". Sageworks, which provides analysis for industries, found that "the industries where U.S. companies with $10 million or less in annual sales have shown the highest...percentage change" included building custom software, building servers for businesses and technical consulting services, while Inc. magazine reports that the current most promising industries for entrepreneurs include apps (currently over $25 Billion) and mobile health (expected to hit $49 Billion by 2020). Technology being paired with entrepreneurship is important, because people who create new businesses are building the future, and if the future is built with technology, then software will have powered it, and software will have eaten the world.

Demand for Programmers in the Workplace

Just over 13 years ago about 50 million people had internet access. The number is currently around 2 billion, a number that is ever increasing. Furthermore, computers could do far less and software related businesses cost far more to run – the cost of running the same businesses is now about a hundredth of the price it was just two decades back. In the Wall Street Journal, Marc Andreessen, co-founder of Netscape, points out how consumer products like books, movies, songs, videogames and photography are all digitalized and their entire industries transformed, along with other businesses such as marketing and telecommunications.

To put it simply, *everything* has become dependent on technology, and in turn massive corporations have become dependent on technically skilled employees with job titles like System Designer, Database Administrator, Network Engineer, Web Developer, Software Engineer...the list goes on. All of this boils down to lines of code, and the need for experts who understand what those lines of code mean is only increasing. Andreessen states that, while he is optimistic about the business environment that the USA provides, "many people in the U.S. and around the world lack the education and skills required to participate in the great new companies coming out of the software revolution. This is a tragedy since every company I work with is absolutely starved for talent. Qualified software engineers, managers, marketers and salespeople in Silicon Valley can rack up dozens of high-paying, high-upside job offers any time they want, while national unemployment and underemployment is sky high...There's no way through this problem other than education, and we have a long way to go." The world needs more programmers.

While the outlook is tough for many jobs, with the current unemployment rate being around 7.5% in the United States and over 12% in Europe, software engineers have a 2.8% unemployment rate. That is not the lowest rate of all careers, and software is not even among the top ten lowest unemployment rates (astronomers and physicists held the first, at 0.3%, followed by biomedical engineers), but it is very low compared to the vast majority of jobs. Furthermore, jobs in computer programming, web development and software engineering are expected to grow by 8-14%, 15-21%, and 23%, respectively, which is considered "above average". That probably has a lot to do with why BusinessWeek, U.S. News, MSN Money, Forbes, CareerBuilder and Huffingtost Post, among other sources, named Software Developers as the #1 job for 2014, and web developers always holding a place in the top ten.

The challenges that hiring managers experience in finding developers are reflected in recent articles titled, "Tech boom! The war for top developer talent" and "Now the Developer is King". Companies lavish their best benefits on software developers, including high salaries ($93,000 is the average), free food and (especially) coffee, health-care, and other perks such as fun work environments with nerf guns, ping-pong tables and more. This is largely because for every ten developer jobs listed, only eight candidates are available to fill the position. The difficulties of hiring a programmer only increase for cities not considered "tech hubs". It took Groupon 18 months to hire the thirty developers needed for their Chicago office. Zach Kaplan, CEO of Investables, says that while his company is at maximum employment for every other role, it desperately needs software developers.

The difficulty also increases as the required skill level for a role increases. Just look at the placement rates for the majority of programming bootcamps. You will be hard pressed to find one with a placement rate below 90% in a three-month span. That is low for a top-tier 4-year degree, let alone a 4-month-or-less programming bootcamp! Companies are willing to train junior developers in their roles to organically grow them into the role they are supposed to fill. Initially, most junior programmers are a drain on their company, but with just a few months of training, a junior developer can become incredibly useful to the company that hired them, which is why startups and mega-corporations alike are now drinking the programming bootcamp Kool-Aid. Most companies are looking for senior developers, but often fill those roles with persons who have only a couple of years (I have heard of less than a year) of experience.

Pair this intense need for software engineers with a shift in the way education is being done and you suddenly have the emergence of new systems, namely, programming bootcamps. With a programming bootcamp, you get developers who are yes, junior, but who typically have an intense drive to learn and who are already working with the hottest technologies in the marketplace. There is, therefore, a delay in all software creation, and that delay is being caused by the incredible and steadily increasing demand for programmers with only a small pool of candidates to choose from.

My Experience

In September 2012 I crossed the pond, changing my home from beside the River Thames to beside the Charles. I had just graduated from college and decided that I wanted to start my own business. I grew up in Connecticut, and Boston is definitely a different space. For one thing, people are very focused all the time, whereas Connecticut has a more laid back, beachy vibe. Furthermore, *everyone* is "wicked smart". I was suddenly exposed to all of these people graduating from top schools who had more masters degrees and PhDs than they could count.

A friend of mine invited me to Venture Cafe in Cambridge, a meeting ground for the entrepreneurial, the technically savvy, and other interesting people. Soon, I was being exposed to new technologies and huge ideas, as well as fully funded companies making a major impact. It was challenging and very motivating, and after some time I was working on my own business venture making a consumer product. At the same time, I began learning to Python. I had started meeting a lot of programmers and companies that were very technical, and I wanted to understand a little bit more about what they were doing. A friend generously offered to give me some lessons. I'm pretty sure my first program was

```
name = raw_input('What is your name?\n')

print 'Hi, %s.' % name
```

Soon after that I was making programs do basic tasks, and writing simple text-based games. It was very exciting! In the meantime, I was working part-time and on my own business. After about nine months of this I decided that my business was not going anywhere fast, and that I needed a change. The business was put "on hold", where it has remained for a while now. I began reading a lot of occupational books and taking job tests to get a sense for what professionals were saying. Interestingly, a few of these resources pointed to programming as a good career choice for me. Funnily enough, the reasons that the tests gave for programming being a good career choice were the very reasons I had decided against programming as my college major, despite the judgments of a test I took my freshman year of college.

For example, creativity. I had always pictured the job description of a programmer as the scene from The Matrix, where Morpheus and Trinity attempt to rescue Neo from his cubicle and life of drudgery. Instead, programming is very creative and can involve a lot of aesthetic design and unique problem solving.

Also, collaboration. Programming is no longer a job for people sitting by themselves, working on stuff quietly. It is a job where creative people work together on cool projects and attempt to build beautiful websites that draw people in, and hardware devices that increase connectivity between our lives and the internet.

My chief concern, however, was that I had only just finished college a year ago and didn't want to have to do it again! Four years is a long time! Also, I didn't want to have to shell out another $100,000+ to get a degree, only to find that the technologies I had learned weren't in demand, or it wasn't a good fit after all. I looked at some community colleges, hoping these might offer direction. The effect was the exact opposite. I spoke with professors and deans at a number of colleges and was told that I would have to complete other prerequisites, most notably, MATH. If there is one class that I am glad I don't have to take anymore it is abstract math. So of course the thought of having to study a lot of math for my programming career sounded awful. Also, I attended one "Intro to Computer Science" course for a day and discovered that more than half the class had dropped out - from 15 students to 6! I started to think that maybe it was just too difficult a switch, and that I needed to have been doing it since infancy.

The same week at a gathering I met a programmer at Intel and told him about the difficulties I was experiencing in pursuing a programming career. He advised me that there were so many learning tools available that I could simply teach myself. I had been teaching myself a little, using online learning tools like Codecademy, but the pace made it expensive in terms of opportunity cost; all of the additional time it would take to teach myself meant that much more time not getting a paycheck. I needed some sort of mentorship learning experience.

My parents were of course aware of my seemingly manic pursuit of this new career choice. My Dad was concerned because most of the "whiz kids" from my high school were now out there doing all of the programming jobs. "Whiz kid": someone who taught themselves to program at a ridiculously early age, built robots for fun, and solved calculus problems before being able to read. It was at around this point in time that I started looking into programming bootcamps. I had read a little bit about these bootcamps, and they seemed to be gaining quite a lot of interest. Websites like CourseReport.com showed that there were nearly 100 programming bootcamps in the world at the time...probably more if you counted bootcamps with multiple locations. The concept was that if you attended these camps, they could take you from not knowing that much about programming to "Junior Developer" level.

This was the incentive I needed. Suddenly the opportunity costs of learning to code were dramatically lowered (assuming that these bootcamps were legit). The camps varied in time and cost, and of course location. I did a lot of research and found that I actually had two camps near me at the time. Factoring in the cost of living, relocation, etc., and that I already had an understanding of the tech community in the Boston area, I decided to apply to camps in the Boston area. Besides, Boston is estimated to be the second largest startup scene in the United States, and among the top 10 largest tech scenes for the entire world, depending on what metrics you use.

After speaking with a couple of programs, I was advised to start programming more. The thinking behind this is simple: try it and see if you like it. I started programming in Ruby using Codecademy, and read through Chris Pine's hands-on book, *Learn to Program*. I learned a lot during those intense weeks, and when I began applying for the camps felt that I might have a good chance of getting in.

I applied to the (at the time) two immersive camps in the Boston area, Launch Academy and General Assembly. Both are well respected camps, Launch Academy having only one location and General Assembly having eight locations. Although both camps state that they you can go in without any experience, you are expected to have been learning programming prior to applying, which shows that you have the drive you will need to succeed. Each camp had me do a test of sorts, which was to go through some curriculum and afterwards build a personal site.

General Assembly had me come in for an interview, but I think that I didn't pass because of the logic based question. It was actually a riddle: you have two candles which can be burned on both ends, each candle burns for sixty minutes exactly, please correctly time forty-five minutes. I was so close, but didn't follow it through to the end, which I think is what probably got me cut. Launch Academy also did not accept me, saying that I was a little bit late applying for that cohort, and that I should apply for the next cohort. Yeah, I got rejected from both.

I stepped back for a moment from the pursuit of programming and started thinking about pursuing an MBA. It seemed as though many students who had pursued an MBA had seen a dramatic increase in their salaries, and an MBA is almost THE thing to do if you're not sure what you are doing with your life. I began studying for the GMAT, the test that most business schools require for entrance. In the meantime, I started delivering pizzas and spent a couple of days a week at an internship with a tech startup.

All of these activities occupied a great deal of time, and I wasn't able to do as much programming as I hoped. Ironically, I discovered an article, albeit an older one, titled "A Smart Investor Would Skip the M.B.A." The article points out that instead of spending the $175,000 on a top MBA you could take that money and invest it on starting a business, or even invest a significantly lower amount on attending a programming bootcamp. Someone with under a year of experience with an MBA will average a $45-$50,000 salary, while a developer emerging from a programming bootcamp will usually make $70,000+ their first year. A 40% higher salary, plus a significantly lower cost (the average programming bootcamp is going to cost about $10,684, while the average MBA will cost $80,000. I cover this subject more in the chapter "MBA vs. DBA (Developer Bootcamp Attainment)", but the idea is simple...MBAs are expensive, DBAs (Developer Bootcamp Attainment) are *relatively* inexpensive.

Yet it wasn't the close analysis of such statistics as these that pushed me onto the world wide web, but a series of *seemingly* unfortunate events. I was working on a random business venture around November and took a significant financial hit, forcing me to begin working a lot more hours. I took a job with Uber as a driver and picked up extra hours at the pizza place where I had been working in the evenings. One snowy January evening I was just finishing my shift at around 11PM and driving home when I suddenly lost control of my car on a turn. For a few seconds I was terrified as I frantically tried to regain control, but to no avail. The car smashed into an illegally parked car on the other side of the street. Luckily I was okay, but I was quite shaken up by the incident. My car was drivable, but looked terrible, so I had to put it in at the repair shop. The car would take nearly two weeks to repair, so I had a lot of free time to think.

I began thinking about what I wanted to do with my life and how I was still stuck doing a job that I had never wanted to do and how I had drifted from my original goal. I looked through my old mail and saw some old email relays that I'd had with the Boston programming bootcamps since after the application had taken place, where they had suggested I reapply but I hadn't because it was bad timing. I sent off an email to Launch

Academy, wondering if their Spring cohort still had spaces. I received a prompt reply telling me to go to the open house that evening. I showed up for the event and then had a Skype interview, in which I was asked to explain what different lines of Ruby code would do. I was easily able to understand the code and describe what it would do, and a few days later I had my official acceptance letter!

It had all come so fast, and I told my boss that I couldn't work there anymore because of school and sent my iPhone back to Uber. My next issue was paying for the camp and pre-work, which I speak more about in the chapter "Paying Your Tuition $$$", and "Prework: What to Expect". The pre-work was very intense. I started off by completing an online Ruby on Rails course, and then got to work going through different reading material and RubyMonk. While there was an even mix of both theoretical and hands-on, that was the only time during the course where I could say that, since the work tended to lean heavily on the hands-on learning experience.

And suddenly it was the night before the program started. I went to Launch Academy to meet the new cohort and we began setting up our new coding environment, mine on a shiny new Mac that I would be renting for the duration of the cohort.

The first day was just a lot of information. We learned a lot about the things we could and couldn't do, and what our time would be spent doing. For Launch, it was as follows:

Alpha: Ruby basics, Object Oriented Programming concepts, HTML and CSS review from pre-work.

Bravo: Databases, Schema Design, SQL, ActiveRecord Models, and Unit Testing with RSpec.

Charlie: Rails Views, Controllers, and Routing. Building complete Rails applications with outside in development.

Delta: Evaluating and incorporating third party libraries. Sending email, queueing, and authentication. Design patterns and more advanced object oriented programming concepts.

Echo: JavaScript with JQuery and unit testing, API integration (OAuth, Scraping with Nokogiri, and consuming JSON api's with Faraday).

Of course, the subject didn't go exactly like that, because each cohort is custom, and so the curriculum is shaped by how the cohort handles the various bits of material.

The first two weeks, or "Alpha", were absolute madness. It was like that song, except the lyrics would be more like "Sit down (code), stand up (code), pass out (code), wake up (code), faded (code), faded (code)"! I can honestly say that I was doing around 14 hours of coding per day, sometimes even more. Usually the day went something like, get in around 8:00AM (some early risers were there at 7:00AM), work on the coding challenge that you had fallen asleep trying to figure out (and woken up thinking about), do "stand up" where you stand up and say in two minutes what you did the day before, learn about something for a short while, pair program on something until lunch, come back from lunch early to figure something out, get another lesson from a mentor, code.

To me, this explanation of my day makes sense, but I imagine that you are wondering why so much of it is spent coding (or maybe you're not...). If you are wondering, then the reason is that programming bootcamps are "hands on". It is all about doing, not about theory. Yes, it is important to learn about things like Document Object Model (DOM) and Client-Server Architecture, topics covered in some depth at most programming bootcamps. Yet sometimes, and often in programming, you don't need to fully understand why something works before you start using it. As crazy as that sounds, it just makes things simpler. You build something, and sort of have to trust that it is going to work, and suddenly you find that it does in fact work! Then you go back and figure out why. That way, you keep going deeper and deeper into new technologies until finally, one day, you reach mastery.

Some members of my cohort might be rolling their eyes while they read this, while others might be nodding their heads in agreement. Everyone has a different style of learning, but there will be many points in a programmer's career where they are taking leaps of faith. You don't know if it will work? Well, try it. The only thing that will break is something that exists in a virtual reality.

So you code and you code. Sometimes it was a struggle where I just couldn't look at another line of code. It is important at those times to simply get up and walk away from the code. Otherwise, you'll end up banging your head on a keyboard and not thinking about how to solve the problem from new angles, while instead you could be taking a leisurely stroll and letting yourself decompress until suddenly the solution will hit you like a lightning bolt! It might sound hard to believe, but the power of your subconscious processing is profound. It was not uncommon for people to cry, and we made our own application to tag people who had fallen asleep during the day (launchersleepy.heroku.com).

Frequently, Friday nights meant partying. A lot of people described it as sort of like being back in college. You're around the same people all the time, to the point where you know who's in the stall next to you because you know what type of shoes everyone wears. People would bring in packs of beer and alcohol, we would hook up Grooveshark to the large speakers and then just dance and laugh.

Monday mornings were not actually that difficult, because over the weekend we were given coding challenges that we needed to solve. Thus, a great deal of my weekend was taken up with writing a program to act like a Cash Register, or writing a website for a kickball team...the list of projects seems endless. Monday mornings were almost just a continuation of the seemingly endless struggle.

I don't mean to paint a bleak picture here. Programming is actually fun. If it wasn't, I wouldn't be doing it. There is nothing like creating something out of nothing. You have an idea? Build it. You don't know how to build it? You pore over StackOverflow, random code blogs, and "The Oracle" (Google) until you've scrapped together enough bits and pieces of code to get the job done. Twice we had "Ship It Saturday", where you show up at 9AM on a Saturday and code until 6PM, and push whatever you built onto a website. So you essentially build a web application in a day.

Prior to programming, I was one of the soft skills people. I mean, I'm good at marketing and talking things up and yes, it is very useful, but it can be very frustrating when you are looking for a job and you realize that your General Studies degree, which didn't come cheap, didn't give you the hard skill-set that will get you a well paid job. At one point, while I was slaving away in the back kitchen of the sandwich place where I worked, I realized that I and all of the other people at the job were not stupid and it wasn't for that reason that we were working for slightly above minimum wage. I mean, one of the employees went to Harvard, one had already graduated from Brown and was going for her masters at Columbia. But we were working these jobs that let our brains die because there had been a number of paths that we could have chosen and we went for one that wouldn't get us to the place we wanted to be.

The most difficult part about having a skill that is not really being sought after is that, even if you do get a job, the likelihood is that it won't be extremely challenging. You might actually feel that it is below your skill set. "You mean to say, I read and re-read half of the Political Science classics so that I could do this menial job every day? I trained my brain for this?" I never have that feeling with programming. The expression "drinking from a fire hose" was frequently used within our cohort, and it was used accurately. So yes, we struggled for ten weeks and put in 14+ hour days, learning Ruby on Rails, Javascript, HTML, CSS, Jquery, SQL, Test Driven Development and a lot more, but our minds grew as did our characters. And, at least for me, I felt as though for once in my life I was learning a skill that would only become increasingly sought after, and for which the learning process was extremely stimulating. At the end of the day, you also get the satisfaction of seeing what you built in action, as though you built a miniature machine, only within a virtual world.

Literally ten weeks of this...700-850 hours over a ten week period, and suddenly you're done. Things don't grind to a halt, however. The day after I finished Launch Academy I submitted my resume to a few recruiting agencies and on a few websites. It was unbelievable. All of the agencies wanted me to go in and speak with them and I was fielding at least 5 random recruiter calls a day (that was on a really low day). It is a strange feeling to go from trying to get a job that will probably only pay $30,000 your first year and having zero success and feeling like scum, to looking for $75k (starting salary) jobs where you're treated like a rock star. The part I like the most is that whereas before I would dress up with a tie for interviews, for developer jobs it is expected that you show up in something like slacks and a long sleeve shirt, maybe with your shirt sleeves rolled. Way more casual and relaxing. Oh, and boat shoes.

The first person got a job within one week of the program ending, and within two weeks about fifteen percent had jobs. Now, writing exactly two months after the end of the program, over fifty percent of my cohort has received job offers, with 100% of people from the previous cohort being hired.

I interviewed with a few companies from the Launch Academy hiring partner pool, but didn't find a good fit, although a number of "Launchers" were hired by those companies. I refined my resume and cover letter with a lot of help from the in-house career expert at Launch Academy, and began hitting the main online job sources. After a week of this, I discovered a site, BostonStartupGuide.com, that listed twenty-seven pages of startups that were hiring in Boston and began sending off cover letters and resumes, trying to cross off one or two pages from the list per day by sending applications to every company that had a position I could fill.

I had some luck and started getting requests to interview, with the infamous "code challenges". The first coding challenge that I got was actually from a Python shop. I had a phone screen with them, and explained that I didn't know much Python but that I was learning, so they gave me a challenge, which was to build a spreadsheet that could calculate all of the values in an array. For example, inputting [2, A1+5, B1 +7] would return [2, 7, 14]. This would need to be possible for infinite rows and columns.

I frantically began working on the challenge, while also interviewing at another company that gave me a coding challenge. This second challenge was to build a simple web page using CSS and JavaScript embedded within a single HTML file which loads something into Google Maps and shows something pulled from an API on animated charts that would update when map points were clicked. The fact that I received both of these requests at the same time was pretty heavy. Fortunately, I had a much longer time to solve the Python problem, so I worked quite intently on the second coding challenge. I was only able to get 80% through the second problem before I ran out of time and had to submit it uncompleted, so they gave me a second challenge! There was also a white boarding session during the interview where I needed to explain how one section of my site sent GET and POST requests to an API, and how the MVCs of the site interacted. This was a more established company, so it made sense that the company had an intense interview process.

Another company that I was interviewing with, a startup, was much more laid back. The company gave me a couple of phone screens and then wanted to interview me in person. Startups start off laid back in the early stages of development, and become more rigorous in their interview process as they age and grow. This is just an observation, and by no means a cover-all.

The first chapter of my own story draws to a close with the exciting news that I have been offered my first position as a developer! Of course, the next chapter of my life will involve a learning curve that is just as steep as I plunge into the world of corporate grade applications.

Three Types of Bootcamps, in Terms of Pricing:

When considering a programming bootcamp, or any form of education, for that matter, one of the most important factors is usually the cost. If you're anything like me, you are wondering to yourself "Where am I going to get the cash to pay for this bootcamp?". While the price range for camps varies from under $5k to $36k+, there are different methods of payment that might help you out, depending on how much (or little) you have available to spend on your bootcamp, and how you feel about debt.

In terms of financing, a number of options exist. I have met people who have done all of the following: credit cards, pay-in-full, peer-to-peer lending/borrowing (P2PL) and investor backing.

Needless to say, there are a lot more risks involved with credit cards and it is probably a good idea to avoid them unless you are really good with handling your finances, or you could end up in a world of debt. If you do decided to use credit cards (you might need multiple cards depending on how much your bootcamp costs), consider getting one that has a 0%

introductory rate. That way, you'll be able to save enough dev cash to pay off your debt before the normal APR kicks in.

Peer-to-peer lending is a relatively new concept, having started in the mid 2000's. It is what it sounds like – borrowing from random people instead of a bank. The advantage over a credit card is that you usually pay a rate lower than half the APR, perhaps 6-10%, which will lessen the financial burden significantly. Investor backing is also new, and similar to P2PL. The only difference is that people lend you money based on your experience, sort of like getting a loan based on your resume.

If you're not in the development world, then you are probably skeptical about whether it is true that someone can get a job after a 10 week programming bootcamp. I mean, who wouldn't be? I can attest to the insanity of the job market right now: every company is always hiring, and of the 5,000 or so people graduating from programming bootcamps a year, most are being hired in less than three months. But don't take my word for it...you can hedge your bets. To hedge your bet is "to leave a means of retreat open." In horse racing, hedging your bet happens when you bet on multiple race outcomes (horses) so as to avoid losing everything in the event that a horse performs poorly. You don't want to risk everything on a horse you're not sure will win.

That's where one type of programming bootcamp makes it easy for you. Many programming bootcamps are so confident that you will find a job upon graduation that they offer job guarantees. One camp states that "the default job-offer guarantee level is $60k+/yr, and top applicants are guaranteed at the $100k+/yr level", and another states that their "graduates are guaranteed an IT job offer within 90 days of graduation or money back". Thus, a person might join a bootcamp that holds to the "no job, no pay" philosophy.

Next on the scale would be joining a bootcamp that offers some sort of kickback upon hiring. Essentially, the bootcamp makes money off of you, the student, by acting as their own recruitment agency. Thus, like a recruiting agency, the bootcamp will be paid a referral fee for every graduate that is hired, and usually some sort of fee for giving employers access to their students, with the potential of finding a good hire. FullStack Academy, in NYC, has a number of hiring partners who, if a student is hired through them, will reimburse the student for a part of his or her tuition. The amount that a student gets back can vary based on the starting salary. What is actually happening is that the employer is first paying the bootcamp their referral fee, often more than 20% of the new employee's salary, and the bootcamp in turn pays a portion of that referral fee, perhaps 35% or so, to the student. Thus, a new hire with a $70,000 salary would receive 35% of $14,000 (20% of their salary), or $4,900. The referral fees vary depending on a lot of factors such as location, average salary, and business model, but the one that was just described is a fairly common scheme.

The final type of bootcamp is your straight-up, pay as you go bootcamp. These bootcamps are actually something of a rarity, the reason being that employers are constantly on the hunt for their next developer hire, and subsequently seek to create partnerships with the programming bootcamps that logistically make sense (usually a combination of what they teach and distance). While these camps are tougher on the student, in that they require the student to shoulder the entire financial burden, there are advantages.

For one thing, the student is less likely to feel constricted geographically to a place near their programming bootcamp. A student who graduates from a bootcamp on the East coast might normally look for jobs on the East coast, and perhaps within 100 miles or less from where he or she graduated. Such geographical constraints can be lessened if the student feels less obliged to look for a job in any particular region. For another thing, the student might be more careful in his or her job selection, instead of jumping on the first offer that comes along. If the student has the opportunity to receive a signing bonus and a decent salary by signing on with a hiring partner, instead of focusing on finding jobs out of all the companies that might be a better fit, it might blind the student as to whether the company is actually a good match. Although these are scenarios that might not apply to most situations, it's worth considering schools where you have to pay your tuition in full.

More important than anything is finding a school that fits you. When applying to bootcamps, I interviewed with five, three of which I didn't want to pursue further, and one that I eliminated from my list after a visit. The first three were sort of "sketchy". They seemed disorganized, and when I discovered that I would be their first cohort, I withdrew.

Programming bootcamps are constantly being shaped and their curriculum is a combination of experiment and requests from employers. And while it might be impossible to avoid being part an early cohort, since programming bootcamps only emerged in late 2011, and while there is no single way to teach a program, a lot of the kinks of a bootcamp are worked out after the first few cohorts. If you're a risk-taker and don't mind being the guinea pig, go for it. If you're highly skeptical and cautious, consider applying to a program that has been around for a little while. General Assembly and Dev Bootcamp might be the largest, most established camps, and were formed in 2011 and 2012, respectively.

However, I think that the size of the organization can be off-putting to some people. While I applied to one of the larger bootcamps, I decided not to attend it because it felt a little too impersonal. Instead, I chose to attend a bootcamp with a single location, because I knew that they had a few cohorts under their belt by the time I started and so had worked out some of the kinks, and that they were more invested in the students.

Furthermore, they were highly adaptive, with the ability to tailor the learning process to our cohort at a moment's notice. Adaptation was very easy with a class of 30, but might be more difficult for a program that has a number of locations and an established system in place.

You also want to consider who started the camp. A lot of programming bootcamps have been started by people trying to "jump on the bandwagon", in an educational movement that is only just starting. For that reason, I think it is important for a programmer to look for a bootcamp that is very passionate about the process of education, and has an excellent staff of not just elite programmers, but educators as well.

Whether the bootcamp you choose is large or small, has a money-back or partial-refund through hiring partners, finding a bootcamp that suits you is an important question, one that you will be able to answer as you speak with and visit the bootcamps that interest you and make sense to attend. If you aren't certain, it is ok to be skeptical. Just don't wait too long.

Interview with David Heinemeier Hansson, creator of Ruby on Rails

Q: How long have you been involved in the programming scene, and how did you first get involved?

A: I've tried to learn how to program since I was 6 years old, sorta on/off. But it wasn't until the late 1990s that I finally succeeded. So I'd say I've been programming "for real" for about 15 years or so. I got started because I wanted to create gaming review websites. Having a concrete mission made it much easier for me to summon the motivation to learn.

Q: Has programming changed a lot since you first became involved, and if so in what ways?

A: I actually don't really think so. The fundamentals are the same as they've ever been. We seem to be moving through cycles like client-heavy or server-heavy and functional vs. object-oriented programming. Plenty of new shiny things, but less fundamental change. I'm still trying to perfect lessons of the past decades.

Q: How would you compare a programming bootcamp to a college education?

A: I haven't actually done a programming bootcamp myself, but I've heard plenty of success stories from others. It all depends on what you want. If you just want to learn how to program and get a job in the industry, then a college education is overkill. But if you want to learn more about the world in general, be a more well-rounded individual, then I think college, or rather, a broad education has much to offer. Whether that offer is currently worth the onerous financial commitments necessary in the US is a different matter.

Q: How do you think online programming bootcamps compare with in-person programming bootcamps?

A: I personally learn best on my own tasked with a project that I really want to see to completion. So for me, it would be online all the way. But different people learn differently, so for others the in-person approach may be exactly what they need to finally make it click. It took me almost 15 years from first seeing programming to finally getting it. Not everyone has that kind of time.

Q: How long do you think that the job market will hold this many programming positions with the currently high pay levels?

A: The billion dollar question, eh? I think it would be foolish to think that the current golden age of the programmer is going to last forever. Eventually the market is going to catch up and better match demand for programmers with supply. Whether that's in 5 or 20 years is more of an open question.

Q: Why is software eating the world?

A: Because, when it works, it makes things faster, cheaper, scalable, repeatable, and all the other siren calls of efficiency and effectiveness. That's incredible value. If you can now do with 2 people what used to take 20, you're freeing up a lot of resources, lowering a lot of costs, and everyone wants that.

Q: Have you seen any changes in the ease of creating a tech startup in recent years?

A: Cost keeps dropping. It's never been cheaper to get started. The main expense is time of the skilled participants doing the work.

Q: Can you comment on the "Internet of Things" and how you foresee it affecting the future world of programming?

A: It's part of the Software Eating the World idea. You can make so many things so much more convenient when they can be programmed.

Interview with General Assembly co-founder Brad Hargreaves

Q: How long have you been involved in the programming scene?

A: I first got started programming my Ti-83 plus when I was bored in math class. I then built a couple of websites and before General Assembly, I got involved in designing a game studio. I have been involved in tech and entrepreneurship scene for a long time. I wanted to bring together the entrepreneurial and the tech scene. When I saw people looking for talent in New York in User Experience, digital marketing, and on the other hand saw a lot of people not enjoying their job, I thought that this was the opportunity. GA started offering a course that would offer a 12 week core skill training for the skills. The rest has been expanding the business.

Q: Has programming changed a lot since you first became involved, and if so in what ways?

A: Everything is now open source technologies – everything we teach, HTML, CSS, Postgres, Ruby, Rails, Ember… is all open source. Ten years ago that wasn't the case. Companies that were teaching were teaching all proprietary Java, Oracle, Cisco networks, which meant that the companies that owned the technology also owned the certifications. The emergence of open source technologies has enables companies to teach technology in a new way because they don't have to pay money to large companies that "own" technologies. Furthermore, open source has enabled new libraries to develop...there is no company that owns JavaScript.

One experience that I had was when I learned Ruby on Rails. I was shocked at how easy it was to get a very simple site up and running. In open source you have such a lot of people contributing to the technology itself. Open source welcomes new participants – it's awesome! At GA we want all of our alumni to not just be taking from the community but also giving back.

Q: Why is the programming bootcamp possible?

A: We get our mission in empowering individuals to do work they love. There is a recent trend of graduates pursuing something where they have a craft. Given the progression of open source technologies it is not nearly as difficult to learn the fundamentals of program as it used to be. It's now possible to get a junior developer position after 12 weeks. Ninety percent of our students are employed after 90 days. It's possible because of frameworks like Rails. The job requirements are not as intense as they used to be. These companies don't need to deal with the billions of pages that Google deals with. There are a lot more use cases that have been created.

Q: How would you compare a programming bootcamp to a college education?

A: It's important to note that 98% of our incoming graduates already have a bachelor's degree. This isn't something where people are choosing either one or the other. We are teaching a very different value proposition. A liberal arts degree will give you cultural contacts: how to think, make you a better citizen. I don't want to say that one is right or wrong...better or worse. I think you are seeing people come to GA instead of a traditional graduate program. Students might be coming to GA instead of going to law school. Clearly being a web/UX designer is a very high growth discipline. I'm specifically talking about the growth of certain disciplines – mobile apps, web development have projected high demand growth rates.

Q: What types of companies are hiring bootcamp graduates?

A: It's all across the board – there is a lot of diversity among companies that are hiring these graduates. As we grow, the companies that are hiring the graduates are getting larger. One of our single largest employers is American Express. They are very excited about our recent graduates, not just for skill but for the creative spirit they bring to our company. We tend to caution our graduates about joining startups, partly because startups don't have the learning resources that our graduates need to grow in their skill. We are seeing our students grow into senior positions within 3 years, especially in larger companies that can allocate mentors to students.

Q: What is the job market like for programmers currently?

A: It is incredibly hot! I mean, as I quoted the stats, over 90% of our graduates have jobs within 90 days. I think that one of the things leading to this is that more and more companies have realized the importance of their online presence. A lot of companies are looking for people to develop their web pages. I believe the trend will continue and increase. You're looking at more and more industries continuing to be transformed by the web. Businesses are realizing that web applications are increasingly important ...businesses need web applications. They are a great way to run the business. You're seeing this need for tech that will last for at least the 5-10 years, likely more. Minus a catastrophic event.

Q: Why is software eating the world?

A: You can look back to the first emergence of the internet in the late 90s when people were just trying to port offline business to online. Until 10 years ago people were figuring out what the net is good/not good at. You're now seeing, for the past 5 years, many of the trends from the 90's reemerge and being effective. The majority of people with broadband are insane. You look at these numbers and you realize why a model like Blue Apron can work. This is one of the things that was tried in the late 90's, only now it is working. You're seeing more and more traditional industries being disrupted. People are saying we need to hire our own dev teams.

Q:Have you seen any changes in the ease of creating a tech startup in recent years?

A: The tools have gotten better across the board. You can look at technologies and see that they improved by an order of magnitude. All of these developer tools have gotten much better since the late 90's. There is more clarity too; companies are no longer as easily taken advantage of by VC's; there are lots of industries, like law, which better understand the needs of startups. It has made the startup ecosystem much more fluid and easier to take part in.

An Interview with Jason Moss, Co-Founder of Metis Programming Bootcamp

Q: How long have you been involved in the programming scene?

A: I'm actually not a programmer. However, I've been working in education for 20-something years.

Q: What motivated you to launch Metis Ruby on Rails bootcamp?

A: The decision to launch Metis was about how to take 80+ years of learning about education and marrying that with world class expertise in software. So we at Kaplan went looking for the perfect partner – and decided that it was Thoughtbot. The programming expertise aspect of the bootcamp comes from Thoughtbot. Instructors are full-time Thoughtbot devs "on sabbatical"...their personal programming experience in the programming field is what helps the students. Thoughtbot has over eleven years of experience building world-class software.

Q: Why is the "Programming Bootcamp" possible?

A: When Kaplan made the decision to launch Metis it was intentional on the topic that it was not simply launching a *programming* bootcamp, but a "new economy skill training". Of course, programming is one of the most highly sought after skills that we teach. However, Metis will also offer training in data science, product design, and digital marketing...these are all skill-sets that have an insane demand. Employers are struggling to fill all of these roles – there are the same pain points for all these roles. For me, the bootcamp has become possible because it provides an accelerated career path to get somebody through the door to these new jobs. It has been demonstrated that in these 10 weeks you can get through to that entry-stage. Bootcamps are not the end of the learning, they are just the first step in the learning. They are being seen and recognized and employers are acknowledging them and validating them. From a macro level we see that the number of people emerging from college with CS degrees can't keep up with programming positions, because software is eating the world.

Q: How would you compare this education model to a college education?

A: The college degree has come under a lot of attack as the student debt figures are rising and the percentage of unemployed graduates is being shared with the public. That said, for me the bootcamp is not a replacement for the college education. Attending a programming bootcamp is a discreet decision...our purpose at Metis is to help get somebody a job. That is a very specific goal, which is not being met by colleges. The decision to attend college is going to be informed by a lot more factors than just getting a job.

Q: What types of companies are hiring bootcamp graduates?

A: There is a natural appeal with startups, but the type of companies hiring bootcamp graduates is broad. There is a little bit of a Catch-22 to this. There is nothing to me that suggest that a bootcamp graduate needs to work at a startup; I think that is just more of a by-product of peoples' natural circles and there is a certain appeal to working at startups. Large companies are starting to work with the programming languages being taught at bootcamps and are increasingly willing to offer apprenticeships, so the net is getting much wider. Kaplan and Metis have a network of employers around the world including not just startups but multinational companies. Metis has every reason to believe that our graduates can work at those larger companies and not just startups. Some of those larger companies may be a *better* fit because the companies will have more resources available.

Q: Can you comment on the emergence of the startup culture? Do you see this tying in to the emergence of programming bootcamps?

A: I would be careful to christen this as a "startup culture moment". If you trace economic activity trends there are lots of sparks over time. Open source software has contributed to the recent spikes in social marketing, just as alternative vehicles for financing are helping with the proliferation of startups, but I don't want to say that we are now in a startup culture whereas before we were not. There is a natural synergy between startups and programming bootcamps because the bootcamps are startups themselves. Also, the bootcamps are teaching languages that are open source. Knowing each other through meetups and networks are contributing to the seeming startup nature of these bootcamps. These companies have the same inherent needs, thinking about UX and design, amassing lots of data about tricks and traffic pattern, which is contributing to why the startups and bootcamps are closely aligned.

Q: How long do you think that the job market will hold this many programming positions with the currently high pay levels?

A: How long will bootcamps be able to fill these positions is another question you should be asking. From my research we are nowhere near the supply and demand lining up - that is going to take a *long* time. What will happen is that certain languages will go out of vogue. For example, Python might fade out while Swift bootcamps might proliferate. There is nowhere near the amount of flow companies need coming out of Computer Science degrees (CS) and bootcamps. 80,000 CS graduates; the numbers are frightfully low. That trend for needing developers is one to stay for a while. I think that bootcamps are a great source for fulfilling that, despite the skepticism. Bootcamps aren't a cure-all. Programmers are going to require a certain level of mentorship and bootcampers will have to have received exceptional training to receive the high pay levels. Our belief at Metis is that there is always going to be a market for top-talent. By partnering with Thoughtbot the feeling was that we would make sure that we are producing exceptional developers.

Interview with Joe Didona of Tech Bootcamps

Q: How long have you been involved in the programming scene?

A: I have been working for 30 years educational startups.

Q: How long do you think that the job market will hold this many programming positions with the currently high pay levels?

A: There are a huge number of tech jobs out there, and the supply of programmers can't catch up with the cycle. This trend will only continue.

When Tech Bootcamps first started our biggest concern was that we couldn't develop *enough* Drupal developers to fill all of the roles. Unless we, (and other educational groups), can put talent out there technologies fizzle out. We literally had one company offer to hire *all* of the graduates from one group we graduated.

Q: Why Drupal?

A: We felt that Drupal could get people to competency quicker, and it's easier from a user standpoint. People know what they can build in a relatively short space of time.

Q: What is the job market like for programmers currently?

A: This was not possible in the past. The old approach was to cram info into people's heads until they graduated college. What has changed is the video technology. There are thousands of videos available for people to learn from. People can study these on their own time in the evenings. Now it's all about project based learning. Huge studies have shown that exams test memory and recall, not ability. The most effective method of learning is the *use* of knowledge. You build stuff, and you simultaneously build a portfolio to show an employer. Furthermore, the open source movement has revolutionized learning. Drupal is open source, so everyone can access it. If you are trying to learn programming, go to meetups and get involved in the programming community. The only way to get in is to teach yourself. Once you're in, you're in.

Q: How can programming bootcamps compare to a college education?

A: What is being taught in computer science degrees is not what businesses need. Colleges have accreditation needs and various certification processes. Open source outpaces those needs. You don't want to go through college for four-plus years, only to find out that the technology you focused on is going out of style.

Top 10 Biggest Tech Startup Scenes in the World

Learning how to program can open doors to places that you had never dreamed of traveling to before, or maybe even working in! There are a number of places that take the cake when it comes to tech scenes.

1) Silicon Valley (Bay Area, California)

Easy. It just *is* the place to be when it comes to technology. Silicon Valley took 40.2% of all of the venture capital deals in the United States in 2012. The *Startup Ecosystem Report 2012* from Project Genome (SER) states that a "pay it forward" mentality is the greatest upside of the Valley, while the insanely competitive search for talent is the greatest downside. Of startups, 73% are making web application, and 17% are making mobile application. In terms of the most used programming languages in the Valley, 28% of startups are using Ruby, 22% are using PHP, 18% are using Python, and 17% are using Java. Interestingly, Silicon Valley is a very spread out area, such that new arrivals are sometimes disillusioned from their originally magical view of the place. Yes, there is hustle going down, but you don't always see it because of the spread. Also, housing is extremely expensive. San Francisco is usually considered the third most expensive place to live in the United States, with average rent set around $3,000 per month for a two bedroom. If you are willing to stomach the potential difficulties of the city and its outrageous prices, expect to make a significantly higher salary than in other cities, get more job offers and work with the latest technologies.

Bootcamps near Silicon Valley:

Angelhack Education, App Academy, Athena Tech Academy, Big Nerd Ranch, CodePath, Coder Camps, Coding House, Dev Bootcamp, Fire Bootcamp, Geekwise Academy, General Assembly, gSchool, Hack Reactor, Hackbright Academy, Hacker Coding Academy, Hackership San Francisco, Koru, MakeGamesWithUs, Makersquare, Marcademy, Mobile Makers Academy, Product School, RocketU, Square Code Camp, tradecraft, Zipfian Academy

2) Tel Aviv, Israel

This city is absolutely exploding with startup energy. Monty Munford of Mashable states that "Everybody seems to be the founder of some new venture, and networking events take place every day, on top of weekend hackathons." Everything is tech and startup and people trying to get something going. As a whole, it is a smaller marketplace than most, but the companies in Tel Aviv are often quickly acquired, making for "faster, frequent exits". The downsides? The "Entrepreneurs in Tel Aviv have a hard time adopting new technology trends, as more than 80 percent of startups use the historically popular programming languages of PHP, C++, Java and .Net," whereas most big tech scenes are adopting Python, Javascript or Ruby, to name a few. For mobile developers, Tel Aviv is paradise. Of startups in Tel Aviv, around 55% are making web applications, while 32% are making mobile application, probably because Tel Aviv has the most advanced mobile market in the Middle East.

Bootcamps near Tel Aviv:

Tel Aviv may not currently have any bootcamps

3) New York City, New York

Although NYC has half the number of startups as Silicon Valley, it "is the second largest ecosystem for software startups in terms of absolute output" and "has been the fastest-growing technology startup ecosystem in the country over the past 10 years". NYC is also prime for startups with a focus on e-commerce, media, fashion and advertising. Shafqat Islam, founder of NewsCred, states, "New York happens to be home to every major industry, making it a huge draw when it comes to running a company...Where else can you have eight meetings in a day with the biggest companies in every sector and still make it home for dinner?" Shafqat also says that while in Silicon Valley almost everyone is in technology, when you leave the office in NYC you find yourself in an extremely diverse crowd of people. Interestingly, NYC is less focused on selling B2B, and more on consumer sales. There is a significant amount of capital going into NYC from venture capitalists; the $2.6 billion invested in 2013 in New York makes its startup ecosystem 87 percent larger than that of Massachusetts according to Tech Crunch, depending on what metrics you use. New York city has drawbacks too. It is the most expensive place to live in America, and, like Tel Aviv, is slow on the technology uptake, and still using more PHP and .NET.

Bootcamps near New York:

App Academy, Byte Academy, Coded,Dev Bootcamp, Flatiron School, Fullstack Academy, General Assembly, Hacker School, HappyFunCorp Technology Academy, Insight, MakeGamesWithUs, Metis, New York Code + Design Academy,

Shillington School, Skaled, Startup Institute, The Data Incubator, TurnToTech

4) Boston, MA

You gotta be wicked smaht if you want to cut it in Boston. While Silicon Valley has 2.5 people with a Masters or PhD to every 1 college dropout, the ratio is 6:1 in Boston. MIT, Harvard, and a lot of other historic schools have led to it receiving the nickname "The Athens of America". Startups in Boston are 24% more likely than those in Silicon to receive funding. This could be because it has an excellent source of venture capitalists, attracted by the significant talent and quality products that those engineers produce. All of these factors combine to make Kendal Square one of the tech epicenters of the world. One entrepreneur states that "Boston is #1 in biotech, #2 in high tech and top 3 in medical devices, energy technology, materials & environmental science, robotics, etc. This is fueled by the large number of top universities and by the diversified New England economy". It is also a great place for startups because it has a number of large startup accelerators, such as MassChallenge, Lab Central, Bolt, and Greentown Labs, as well as the push to open up the Innovation District near Boston

Seaport to more startups. For that reason, tech provided more jobs in MA in 2013 than any other industry. To quote the Boston Globe, "These are boom times for the innovation economy in Boston and Cambridge".

Bootcamps near Boston:

General Assembly, Metis, Startup Institute, Tech Bootcamps,Launch Academy

5) London, England

London, dubbed "Tech City", where tech is primarily located around Old Street and Stratford, its center being the Shoreditch area, is a hub of startup activity. In fact, "London has burst onto the scene and has become the most successful Startup Ecosystem in Europe, producing the largest output of startups in the European Union by far." Part of its growing success is that American companies often choose London as a starting point for expanding their market into Europe. There are also a number of startup incubators, notably BBC labs and TechStars.

One benefit of starting up in London is that the UK has made it incredibly easy to officially *start* a business. As one entrepreneur notes, "a company can be registered online within 5 minutes and VAT registration takes another 10 minutes. Renting traditional offices can be expensive but with the prevalence of shared work spaces and tech incubators, it's not hard to find a reasonably priced office." A downfall of the London startup scene is that it is more risk-averse than other markets. However, founders have an excellent mindset, with their focus on changing the world. London also uses older technologies more often, such as PHP, used by 50% of companies, as opposed to Ruby, used by 13% of London tech startups, and Python (8%).

Bootcamps near London:

Coderoute, General Assembly, Makers Academy, Science to Data Science, Steer, Supacoderz, We Got Coders, Startup Institute

6) Toronto, Canada

For a relatively small city, with a population of just under six million, Toronto is growing rapidly in the startup scene. The scene has some interesting demographics, with a 1 to 1.4 ratio of company founders being college dropouts to those with masters and PHD's. Findings show that Canadian entrepreneurs have a near ideal startup mentality.

In fact, large tech companies (all the big ones) have been attracted to the scene as has been talent from other areas of the globe. When TechCrunch visited the MaRS Discovery Project, a 700,000 square foot tech and startup environment, they realized what a burgeoning place Toronto is on the techdar. With 30% of Canada's information and communications tech workforce, Toronto "enjoys an exceptional 95.9 per cent employment rate and an average annual wage of $64,725". The founder of one company points out that if "you think about Facebook, Google, all of the big Valley companies – most of them were started out of the universities. It's the talent from the engineering schools that fuelled the tech scene in Silicon Valley". With nearby schools like University of Toronto, Queen's University and University of Waterloo, Toronto is similar to Boston in that it has a nearby talent pool from which to fish. Not to flog a dead horse, but Toronto just might deserve the nickname "Silicon Valley North".

An unfortunate reality for the Toronto market is that it employs a lower number of people per funding stage than many other startup scenes, largely because the scene is somewhat lacking in funding/investors at its various levels. As one more pessimistic blogger notes, "Toronto is a strange startup town. Sort of like New York, but without any of the VC money. High cost of living; without the salaries that come with it." However, as the growth trend continues, these negative factors will almost certainly balance out.

Bootcamps near Toronto:

Bitmaker Labs, BrainStation, hackerYou

7) Vancouver

Vancouver consistently wins awards for "Best City in the World" and "Most Beautiful City in the World", and their license plates read "Beautiful British Columbia". The tech scene is also exploding with growth. Entrepreneurs here are extremely hard working, and seriously underfunded. The focus is very much on mobile, with 30% focusing on mobile compared with 17% in Silicon Valley, and only 58% focusing on web startups compared with 73% in SV. A recent article notes that the "old mix of mining stock promoters taking liquid lunches at the Marble Arch and tattooed gym rats of dubious profession in Hummers is being replaced by skinny jeans and Macbook Pros...at restaurants here you are as likely to encounter a terrine of duck foie gras or pan fried veal sweetbreads as you are a Subway sandwich. The rents, while still cheaper than tony Yaletown, are climbing". Perhaps the high rent is worth it, with average salaries among programmers on the rise. One startup notes the difficulties of hiring, since "Vancouver's tech talent crunch is no longer looming – it is already here, and it is particularly pronounced for small, low-profile tech startups that are competing" with much larger companies, and even the larger companies are noticing the shortages of talent. A lot of tech accelerators have a presence in the city as well, including GrowLab, InstituteB, Wavefront and VentureLabs. For a very engaging, visual representation of Vancouver's tech history and scene, it is worth visiting www.visualcapitalist.com/tech-vancouver-timeline-infographic/ .

Bootcamps near Vancouver:

Lighthouse Labs, CodeCore Bootcamp

8) Paris

Paris has seen massive transformation in its tech scene, with major business schools offering accelerators, and a shift in the mindset of many Gen-Xers towards a more entrepreneurial spirit. Many have noticed this shift in France towards technology, with billionaire Xavier Niel sponsoring 1,000 students to join a three year intensive program where they will intensely learn computer science, as well creating a startup incubator for 1,000 startups. There are a number of other startup accelerators to choose from, such as Le Camping and TheFamily. Paris is also home to some of the world's greatest universities, such as H.E.C., École Normale Supérieure, École Polytechnique and EMLYON, leading to the 97% of founders who have a Masters or PhD versus the 42% in Silicon Valley.

Similar to many of the cities in the top twenty, the main problem with the Parisian tech scene is a lack of funding, and Forbes estimates that "a start-up that raises $10 million for a Series A in the US, would raise $500,000 in France". Accordingly, "Paris startups raise 95% less capital in stage 3 (Efficiency Stage) and 91% less capital in stage 4 (Scale Stage) than SV startups", and are mainly self-funded, or funded by family members or incubators.

This lack of funding has caused Startup Genome to note that one of Paris's major setbacks is its failure to attract talent from abroad. This is of huge significance, since "first-generation immigrants were on the founding teams of roughly 52% of all tech companies in Silicon Valley", and at least one out of six entrepreneurs in America is a first-generation immigrant. One Parisian entrepreneur notes that the mindset is good, but still needs to be pushed to expect greater things when he states that "we still need VCs who will tell entrepreneurs [that they want them to] disrupt the whole industry and build the next Groupon, the next Foursquare, the next Facebook". Furthermore, Paris is behind on technology, with few entrepreneurs using Ruby and even fewer using Python (the majority are using .NET, followed by C++ and PHP). One TechCrunch writer optimistically points to Paris' future, saying it is where NYC was a few years ago, and just needs time to grow.

Bootcamps near Paris:

Paris may not currently have any bootcamps

9) Sydney

Sydney and Melbourne combined account for over three quarters of tech startup founders and venture capitalists for all of Australia. There have been huge amounts of government spending, matched by industry leaders, in order to attract talent and create jobs, but the technological advancement has come largely from the inside. As a result, "Sydney will be Australia's global gateway for innovative new industries that create tens of thousands of new high-tech, high paid jobs of the future and add billions to export revenues with three new industry partnerships in the growth sectors of transport, financial services and ICT." The optimism is well founded.

A number of startups from Sydney have gone public, and there are many with huge user bases. Furthermore, Australia has many of the building blocks of a tech hub, such as incubators and co-working spaces, as well as meetups. Even more importantly, Australia ranked first out of the 20 top tech hubs for "trendsetter index", a metric that Project Genome states might be the "leading indicator of the future success of a startup ecosystem", that "measures how quickly a startup ecosystem adopts new technologies, management processes, and business models.

Where startup ecosystems that stay on the cutting edge are expected to perform better over time," Australian techies seem to innovate quickly. The New South Wales government notes that Australian techies have an "openness and flexibility" which "can mean that the product is wildly innovative and in a position to catalyze a whole market, causing a paradigm shift" and many businesses to set up in Australia, including "Amazon Web Services, Williams Sonoma, Havas, Twitter and many more."

Furthermore, Sydney has one of the strongest talent pools out of all the tech hubs. There are a number of excellent universities in Sydney, which partly explains why entrepreneurs in Sydney have similar levels of Masters and PhD education. The main complaints about the city are, of course, a lack of funding. One venture capitalist comments that "We have an urgent need for capital in the Efficiency stage. At this stage of the startup lifecycle there is practically no investment happening and companies are failing for the wrong reasons. We are seeing signs of life here but have a long way to go." Ideally, Sydney will see an increase in funding as it continues to output highly creative products.

As a city, it is one of the best places to live, in terms of lifestyle, and worst in terms of cost. Consumer prices are about 15% higher than in San Francisco, and many of the other costs are comparable. The only big saver is on rent, which is made up by the freakishly high costs of utilities. While the minimum wage in Sydney is currently over $16.85 per hour, a software engineering position will on average pay a lower rate ($73850) than a job in Boston ($80,826) or San Francisco ($100,641). Either way, as the locals of Sydney would put it, "No worries, she'll be alright".

Bootcamps near Sydney:

Coder Factory, Fire Bootcamp, General Assembly, Polycademy, Sydney Dev Camp

10) Sao Paulo

One of the largest cities in the world, a country that is booming with activity and boasting the 2014 World Cup and 2016 Summer Olympics, is also a top ten tech hub. Sao Paulo hosts many of the coolest, highest tech companies in the world, including Google, AirBnB, and BuscaPe, and is fostering a large number of startups through co-working spaces such as Startup Mansion and Pto de Contato.

Furthermore, one founder states that "since it's the economy hub, it has a larger supply of entrepreneurs, MBAs, financial analysts, and investors". Whether it's the weather or the huge economic growth that Sao Paolo is seeing, something about the market is attracting the right talent and the right investment. After seeing huge "economic change with the rise of the lower income classes that generates a massive consumer market that will evolve over the next 10 or 20 year", Sao Paulo has the potential to continue its high growth rate. This means more business opportunity for technology startups.

Sao Paulo's main issue lies in its scattered "ecosystem". You don't have the same sense of ingenuity and creation and the feeling that all of the parts and businesses are interconnected. In fact, commuting is a huge issue in Brazil and can take hours because of the congestion.

Bootcamps near Sao Paulo:

Sao Paulo may not currently have any bootcamps

MBA vs. PROGRAMMING BOOTCAMP

This section will focus on comparing an MBA with a programming bootcamp, because in terms of educational career paths, one of the most common decisions people make is to attain the illustrious "MBA", with the idea that it will add prestige to the resume and dollars to the bank account. Of the 12% of people who get their masters, a major portion is MBAs. Also, the concept of an MBA is probably the closest concept to a programming bootcamp in terms of attendee expectations. For example, of those who received MBAs in 2012, 64% used the degree to switch careers. The same could be said of the vast majority of those attending programming bootcamps.

A recent article titled "Grad School: Still Worth the Money?" focuses on the economics of attending grad school, especially the return on investment. The article points out that the cost of the MBA is not just the roughly 100k average price tag of attending the University, but also the 45k a year (average salary of someone who just completed undergrad) over two years, bringing the average MBA cost to a whopping $190k, plus the yearly interest.

While an MBA, at least from one of the top 50 schools, will pay off dividends over a lifetime, bringing the initial tuition to only a small percentage of the total earned, a $200k debt is an incredible burden to have, and one that will take years to pay off. Thus, despite its paying dividends over a lifetime, an MBA's immediate benefits are difficult to experience. For many it is a frightening sum, one too great to risk. Another article shows that the average starting salary of someone with an MBA "with less than one year of experience was $46,630 ". Of course, it often depends on the school, with an Ivy League or other top school graduate having a substantially higher salary than a graduate from a lesser known school; for example, "graduates of the 57 top MBA programs will earn approximately $2.4 million in base pay and bonuses over a 20-year career", and graduates from Stanford, Harvard or Wharton average around $3.5 million over 20 years in base pay.

The effect of education on earning potential is huge. Lost time accounts for approximately half of the cost, which is one of the upsides of a programming bootcamp. The cost is usually around $12,000 for a 12 week course. That's half the weekly cost of the average MBA (200k over two years = 2k per week), and will land the average person a salary that is $20,000 higher in the first year. Furthermore, when you compare the employment statistics to those of programming bootcamps, you find a nearly identical hiring rate. For example, 91% of MBA graduates from

Yale were offered jobs within 3 months, a rate that is the same and often higher for programming bootcamps, such as Launch Academy (96%) and App Academy (95%).

One study by the US government showing the lifetime earning potential by degree, ranks engineering and architecture highest, with $3.4 million per year earning potential, followed by computer and math at $3.2 million. Fourth on the list are business and finance with an average of $2.7 million per year earning potential. That said, combining two majors, such as management and engineering or business and engineering resulted in very high lifetime income of $4.1 and $3.5 million, respectively. The Hamilton Project's chart of median lifetime earnings by college major also lists computer engineering as the fourth and computer science as the tenth highest earnings by college major.

An MBA can be an excellent investment, depending on where you go and/or what you do with it. Graduating from a programming bootcamp, by comparison, is also an excellent investment and will get someone an almost equally well-paying job in the same amount of time or sooner, and at a fraction of the financial and time risk. Neither an MBA nor programming bootcamp can be "better", because the question is ultimately

"What are your goals?" and "How long are you willing to wait to achieve those goals?"

If your goal is to create a business, in the past you might have focused on an MBA. Yet as the sands have changed in the tech industry, the tools of business creation have become incredibly easy and cheap, such that anyone can create a business in relatively short amount of time. This has lead to a proliferation of technical founders and co-founders, who are often equally if not more valuable to the creation of a successful company. The network of an MBA is notoriously valuable, but so is the network of a computer programmer, who will usually meet both business and technical types at the various meetups and watering holes of the tech world. And as startups get funded, there has been a spread in the knowledge of the business side of technology, to contradict the cliché of a techie with no people skills who knows nothing of the business process. Of course, MBAs have a huge amount of depth to their knowledge, which is why they make large salaries. Yet, for certain avenues, such as startups, an MBA graduate is not necessarily better suited than a technical person.

Therefore, there are pros and cons of MBAs and programming bootcamps, in terms of cost, goals and outcomes. Yet there need not be any dividing line. An MBA graduate can attend a programming bootcamp, or a programmer can get an MBA, and both would benefit from the educational experience. Yet there is no comparison between the time it takes to get an MBA and the time it takes to attend a programming bootcamp. The bootcamp gets you there faster. Someone who is considering both avenues should take into account the long-term as well as short term benefits to each avenue, and more importantly consider what they enjoy. This well rounded approach will certainly pay dividends in the future and lead to a more fulfilling career.

Reading List for the Self-Taught Software Engineer

General:

- Pragmatic Thinking and Learning

Ruby on Rails:

-Programming Ruby 1.9 (3rd edition): The Pragmatic Programmers' Guide

- Agile Web Development with Rails 4

- RUBY ON RAILS TUTORIAL: Learn Rails by Example

- Learn Ruby the Hard Way

- Learn to Program, Chris Pine

- The Well-Grounded Rubyist

SQL :

- Beginning Database Design

JavaScript:

- Eloquent JavaScript

Node.js:

- Practical Node.js

- The Node Beginner Book

A Classic C.S. Degree from Harvard:

A basic computer science degree from Harvard requires twelve "half-courses" in computer science theory, math, and software. The math courses generally involve linear algebra (logic) and calculus.

1: Math 1a: Introduction to Calculus

2: Math 1b: Calculus, Series, and Differential Equations

3: AM21b: Mathematical Methods in the Sciences

4: CS20: Discrete Mathematics for Computer Science

5: CS50: Introduction to Computer Science 1

6: CS61: Systems Programming and Machine Organization

7: CS109: Data Science

8: CS121: Introduction to Formal Systems and Computation

9: CS124: Data Structures and Algorithms

10: CS141: Computing Hardware

11: CS171: Visualization

12: CS179: Design of Usable Interactive Systems

Your First Programs: Ideas

1: **FizzBuzz**: Write a program that prints the digits 1 to 100, except that the program should print "Fizz" for multiples of three and "Buzz" for multiples of five. For any number that is a multiple of three *and* five print "FizzBuzz".

2: **Fibonacci Sequence**: In the sequence 0, 1, 1, 2, 3, 5, 8, 13, 21, 34, 55..., each of the numbers comes from adding itself to the previous number. Write a program that takes in a number and returns the Fibonacci number that occurs at that point in the sequence. For example, the 7th Fibonacci number would be 8, and the 10th Fibonacci number would be 34.

3: **Spreadsheet Calculator**: Given an array of numbers, in the form of a spreadsheet, such as

mini_spreadsheet = [

    [1, "=A1+1", "=A1 + B1"],

    ["=B1", "3", "=C1 + B2"],

    ["A1", "=B1+A1", "C2 + B3", "= A3 + B3 + C3"]

]

, where A, B, C... are the columns and 1, 2, 3... are the rows, write a program that will correctly compute all of the cell solutions.

4: **Secret Santa**: Takes in a list of people (even numbered list) and assigns them a single person that they need to give a gift to (exactly one gift given/received by each person).

5: **Rock, Paper, Scissors**: Build a program that competes at Rock, Paper, Scissors (your favorite childhood game) against you. Simple enough - rock beats scissors, scissors beats paper, paper beats rock. For increased difficulty, the program should be built such that the more you play against it, the more often it beats you, based on your previous techniques and choices. Essentially, it learns and improves.

6: **Factorial**: The factorial of 4 is equal to 4*3*2*1, or 24. The factorial of 6 is equal to 620 (6*5*4*3*2*1). Write a program that will calculate the factorial of any number that you give it.

7: **Roman Numerals**: A program that converts numbers to Roman Numerals. In Roman Numerals, I = 1, V = 5, X = 10, L = 50, C = 100, D = 500, M = 1000. You combine letters by listing them largest to smallest from left to right, which adds the letters (XI = 11, XXV = 25). When you list a smaller letter before a larger letter, the value of the smaller number is subtracted from the larger number (CD = 400, LM = 950).

8: **Dice Roll**: Write a program that takes in the number of times that you want to roll a dice, which returns a count of all of the returned numbers from each roll. For example, If you gave the program three rolls, the output might be 6: One, 5: Zero, 4: Zero, 3: Two, 2: Zero, 1: Zero, indicated that your program rolled a 6 once and a 2 twice.

9: **Phone Typing**: Based on the old-school cell phones that had buttons, where each button represented a few numbers, (2 = abc, 3 = def, 4 = ghi, 5 = jkl, 6 = mno, 7 = pqrs, 8 = tuv, 9 = wxyz, 0 = space), write a program that will convert a text-message into a list of what buttons to push and how many times. For example, "hello you" would return:

h – press four 2 times

e – press three 2 times

l – press five 3 times

l – press five 3 times

o – press six 3 times

space – press 0

y – press nine 3 times

o – press six 3 times

u – press eight 2 times

10: **Most Common Words**: This program should take two parameters. The first is a string of words, and the second is a number. Given this sentence, paragraph, or other random string of words, the program should return an array containing the number of most commonly occurring words which you specified as one of your parameters, listed left to right in terms of most common (leftmost) to least common (rightmost). If there is a tie for the most common words, the program should return the word that is alphabetically sooner (a before z).

Works Cited

Preface:

Bridgewater, Adrian. "Apple Swift New IOS and OS X

    Programming Language." *Push Cool*. Dr. Dobb's All, 10 June

    2014. Web. 22 Oct. 2014.

    <http://www.tuicool.com/articles/bANF7z>.

"Ever Higher Society, Ever Harder to Ascend." *The Economist*.

    The Economist Newspaper, 1 Jan. 2005. Web. 22 Oct. 2014.

    <http://www.economist.com/node/3518560>.

"FACT SHEET on the President's Plan to Make College More

    Affordable: A Better Bargain for the Middle Class." *The White*

    House. The White House, 22 Aug. 2013. Web. 22 Oct. 2014.

    <http://www.whitehouse.gov/the-press-

    office/2013/08/22/fact-sheet-president-s-plan-make-

    college-more-affordable-better-bargain->.

"Henry Ford's $5-a-Day Revolution." *Ford Go Further*. Ford News

    Center. *n.d.* Web. 22 Oct. 2014.

<http://corporate.ford.com/news-center/press-releases-detail/677-5-dollar-a-day>.

"Meritocracy." *Dictionary.com Unabridged*. Random House, Inc. *n.d*. Web. 22 Oct. 2014. <http://dictionary.reference.com/browse/meritocracy>.

"Meritocracy." *Merriam-Webster*. Merriam-Webster. *n.d*. Web. 22 Oct. 2014. <http://www.merriam-webster.com/dictionary/meritocracy>.

"Meritocracy." *Wikipedia*. Wikimedia Foundation, 21 Oct. 2014. Web. 22 Oct. 2014. <http://en.wikipedia.org/wiki/Meritocracy>.

"Money Can Buy Happiness." *The Economist*. The Economist Newspaper, 2 May 2013. Web. 22 Oct. 2014. <http://www.economist.com/blogs/graphicdetail/2013/05/daily-chart-0>.

Savitz, Eric. "The War For Tech Talent: Genius Is Not

Enough." *Forbes*. Forbes Magazine, 23 Feb. 2012. Web. 22

Oct. 2014.

<http://www.forbes.com/sites/ciocentral/2012/02/23/the-

war-for-tech-talent-genius-is-not-enough/>.

Three Types...

Eggleston, Liz. "Course Report Bootcamp Graduate

Demographics & Outcomes Study." *Course Report*. Course

Guide Inc., *n.d.* Web. 22 Oct. 2014.

<https://www.coursereport.com/resources/course-report-

bootcamp-graduate-demographics-outcomes-study>.

"Hedge Your Bets." *Hedge Your Bets*. www.phrases.org.uk, *n.d.*

Web. 22 Oct. 2014.

<http://www.phrases.org.uk/meanings/hedge-your-

bets.html>.

Why Software...

Anders, George. "How to Launch a Billion Dollar Startup on a

Shoestring." *Forbes*. Forbes Magazine, 2 May 2012. Web. 22

Oct. 2014.

<http://www.forbes.com/sites/georgeanders/2012/05/02/t

hrifty-startup/>.

Andreessen, Marc. "Why Software Is Eating The World." *The*

Wall Street Journal. Dow Jones & Company, 20 Aug. 2011.

Web. 22 Oct. 2014.

<http://online.wsj.com/news/articles/SB1000142405311190

3480904576512250915629460>.

Ante, Spencer, and Jessica Lessin. "Apps Rocket Toward $25

Billion in Sales." *The Wall Street Journal*. Dow Jones &

Company, 4 Mar. 2013. Web. 22 Oct. 2014.

<http://online.wsj.com/news/articles/SB1000142412788732

329370457833440153421787B>.

Beim, Nick. "The Rise And Future Of The New York Startup

Ecosystem |."*TechCrunch*. www.techcrunch.com, 28 Feb.

2014. Web. 22 Oct. 2014.

<http://techcrunch.com/2014/02/28/the-rise-and-future-of-

the-new-york-startup-ecosystem/>.

Beliak, Julia. "Is Technology Making Our Lives Easier... Or Just

Adding More Stress?" *The Huffington Post*.

TheHuffingtonPost.com, 21 Oct. 2013. Web. 22 Oct. 2014.

<http://www.huffingtonpost.com/julia-beliak/womens-

forum-2013_b_4138876.html>.

Bersin, Josh. "The Software Economy: Why Software Jobs Are

Taking Over." *Forbes*. Forbes.com, 5 Aug. 2013. Web. 22 Oct.

2014.

<http://www.forbes.com/sites/joshbersin/2013/08/05/the-

software-economy-why-software-jobs-are-taking-over/>.

Clifford, Catherine. "The 10 Fastest-Growing Industries for Small

Business." *Entrepreneur*. www.entrepreneur.com, 2 Apr.

2013. Web. 22 Oct. 2014.

<http://www.entrepreneur.com/article/226256>.

Coleridge, Chris. "MSING016: Strategy for High-Tech

Ventures." *UCL MANAGEMENT SCIENCE AND INNOVATION*.

www.msi.ucl.ac.uk, *n.d.* Web. 22 Oct. 2014.

<https://www.msi.ucl.ac.uk/modules/msing016-strategy-

high-tech-ventures>.

Denning, Steve. "Why Software Is Eating The World." *Forbes*.

Forbes Magazine, 11 Apr. 2014. Web. 22 Oct. 2014.

<http://www.forbes.com/sites/stevedenning/2014/04/11/w

hy-software-is-eating-the-world/>.

Hendricks, Drew. "5 Most Promising Industries for

Entrepreneurs in 2014." *Inc*. www.inc.com, *n.d.* Web. 22 Oct.

2014. <http://www.inc.com/drew-hendricks/5-most-

promising-industries-for-entrepreneurs-in-2014.html>.

Hull, Patrick. "Top 5 Industries for Entrepreneurs." *Forbes*.

    Forbes Magazine, 12 Sept. 2013. Web. 22 Oct. 2014.

    <http://www.forbes.com/sites/patrickhull/2013/09/12/top-

    5-industries-for-entrepreneurs/>.

Jackson, Joab. "The Mainframe Turns 50, Or, Why the IBM

    System/360 Launch Was the Dawn of Enterprise

    IT." *PCWorld*. PCWorld.com, 7 Apr. 2014. Web. 22 Oct. 2014.

    <http://www.pcworld.com/article/2140220/the-mainframe-

    turns-50-or-why-the-ibm-system360-launch-was-the-dawn-

    of-enterprise-it.html>.

Lajoie, Marc, and Nick Shearman. "What Is Alibaba?" *Wall Street

    Journal*. Dow Jones & Company, *n.d.* Web. 22 Oct. 2014.

    <http://projects.wsj.com/alibaba/>.

Lauricella, Tom, Scott Patterson, and Jenny Strasburg.

    "Electronic Trading Glitches Hit Market." *The Wall Street

    Journal*. Dow Jones & Company, 1 Aug. 2012. Web. 22 Oct.

    2014.

<http://online.wsj.com/news/articles/SB1000087239639044

3687504577563001717194274>.

Pai, Aditi. "Global Mobile Health Market to Grow to $49B by

2020." *Mobihealthnews*. http://mobihealthnews.com, 5 Mar.

2014. Web. 22 Oct. 2014.

<http://mobihealthnews.com/30616/global-mobile-health-

market-to-grow-to-49b-by-2020/>.

Pepitone, Julianne. "Facebook IPO: What Went

Wrong?" *CNNMoney*. Cable News Network, 23 May 2012.

Web. 22 Oct. 2014.

<http://money.cnn.com/2012/05/23/technology/facebook-

ipo-what-went-wrong/>.

"Startup Ecosystem Report 2012: Part One." *Multisite -Blog*.

Telefonica Digital, *n.d*. Web. 22 Oct. 2014. <http://multisite-

blog.digital.telefonica.com.s3.amazonaws.com/wp-

content/uploads/2013/01/Startup-Eco_14012013.pdf>.

Sidhu, Ikhlaq. "An Introduction to Technology Entrepreneurship:

Berkeley Method of Entrepreneurship." *Berkeley*.

http://cet.berkeley.edu, *n.d.* Web. 22 Oct. 2014.

<http://cet.berkeley.edu/wp-content/uploads/BME-

Concepts-F20141.pdf>.

Wilson, Jacque. "Your Smartphone Is a Pain in the Neck." *CNN*

Health. Cable News Network, 20 Sept. 2012. Web. 22 Oct.

2014. <http://www.cnn.com/2012/09/20/health/mobile-

society-neck-pain/>.

Interview by Joe Didona

Anderson, Chris. *Free: How Today's Smartest Businesses Profit*

by Giving Something for Nothing. London: Random House

Business, 2010. Print.

Anderson, Chris. "Free! Why $0.00 Is the Future of

Business." *WIRED*. WIRED.com, 25 Feb. 2008. Web. 22 Oct.

2014. <http://archive.wired.com/techbiz/it/magazine/16-03/ff_free?currentPage=all>.

"Joe Didonato: Founder, CEO of TechBootCamps." *TECHBOOTCAMPS*. TECHBOOTCAMPS.com, *n.d*. Web. 22 Oct. 2014. <http://www.techbootcamps.com/team/joe-didonato#sthash.9BGXt0C3.dpuf>.

Price, David. *Open*. Crux, 2013. Print.

Top 12 Biggest Cities...

"5 Reasons Toronto Is Inching towards Becoming 'Silicon Valley North'"*YongeStreet*. http://www.yongestreetmedia.ca/,18 Sept. 2013. Web. 24 Oct. 2014. <http://www.yongestreetmedia.ca/inthenews/siliconnorth091813.aspx>.

"12 REASONS TO JOIN VANCOUVER'S TECH REVOLUTION IN
2014." *Vancity Buzz*. www.vancitybuzz.com,14 Jan. 2014.
Web. 22 Oct. 2014.

"50 Reasons Why Sydney Is the World's Greatest City." *CNN
Travel*. Cable News Network, 6 Apr. 2011. Web. 24 Oct.
2014. <http://travel.cnn.com/sydney/life/50-reasons-why-
sydneys-best-city-world-291350>.

"2012 Venture-capital Deals, by City." *Crain's New York
Business*. http://mycrains.crainsnewyork.com, *n.d.* Web. 24
Oct. 2014. <http://mycrains.crainsnewyork.com/stats-and-
the-city/2013/technology/2012-venture-capital-deals-by-
city>.

"Average Salary for Location: Boston, Massachusetts." *Payscale.
Human. Capital.* www.payscale.com, 18 Oct. 2014. Web. 22
Oct. 2014.
<http://www.payscale.com/research/US/Location=Boston-
MA/Salary>.

"Average Salary for Location: Sydney, New South

Wales." *Payscale. Human. Capital.* www.payscale.com, 18

Oct. 2014. Web. 22 Oct. 2014.

<http://www.payscale.com/research/AU/Location=Sydney-

New-South-Wales/Salary>.

Aysan, Zach. "Toronto: A Strange Startup Town." *Medium*.

www.medium.com, 14 Aug. 2013. Web. 24 Oct. 2014.

<https://medium.com/toronto-tech/toronto-d0ea5da434e>.

Beim, Nick. "The Rise And Future Of The New York Startup

Ecosystem |." *TechCrunch*. www.techcrunch.com, 28 Feb.

2014. Web. 24 Oct. 2014.

<http://techcrunch.com/2014/02/28/the-rise-and-future-of-

the-new-york-startup-ecosystem/>.

Bennet, Nelson. "Tech Talent Crunch Hitting Vancouver

Startups." *Business In Vancouver*. www.biv.com, 5 Nov.

2013. Web. 22 Oct. 2014.

<http://www.biv.com/article/20131105/BIV0112/31105997
9/-1/biv/tech-talent-crunch-hitting-vancouver-startups>.

Benson, Simon. "A $200 Million Plan for the Sydney of the
Future." *The Daily Telegraph*.
http://www.dailytelegraph.com.au, 28 Aug. 2013. Web. 22
Oct. 2014. <http://www.dailytelegraph.com.au/news/nsw/a-
200-million-plan-for-the-sydney-of-the-future/story-
fni0cx12-
1226705329953?nk=099954b65984b00751bd810d06f2d8de
>.

"Best Student Cities in the USA." *Careers 360 Study Abroad*.
http://www.studyabroad.careers360.com/, 30 Apr. 2014.
Web. 24 Oct. 2014.
<http://www.studyabroad.careers360.com/best-student-
cities-usa>.

Brazil Hacks. www.brazilhacks.com, *n.d*. Web. 24 Oct. 2014.
<http://brazilhacks.com/>.

Bogart, Nicole. "Silicon Valley-based Entrepreneurs Return to
    Toronto to Invest in Rich Talent Pool." *Global News*.
    http://globalnews.ca/, 6 Sept. 2013. Web. 24 Oct. 2014.
    <http://globalnews.ca/news/824243/silicon-valley-based-
    entrepreneurs-return-to-toronto-to-invest-in-rich-talent-
    pool/>.

Caldwell, Vanessa. "Tech and the City: Toronto Is Canada's High-
    tech Hub." *MaRS*. http://www.marsdd.com/, 13 Apr. 2011.
    Web. 24 Oct. 2014. <http://www.marsdd.com/news-and-
    insights/tech-and-the-city-toronto-is-canadas-high-tech-
    hub/>.

"Canada's High-Tech Hub: Toranto." http://www1.toronto.ca,
    n.d. Web. 24 Oct. 2014.
    <http://www1.toronto.ca/static_files/economic_developme
    nt_and_culture/docs/Sectors_Reports/Canada_High_Tech_H
    ub_lores_.pdf>

Coleman, Alison. "Entrepreneur: The French Do Have A Word
    For It." *Forbes*. Forbes Magazine, 14 Feb. 2014. Web. 24 Oct.
    2014.
    <http://www.forbes.com/sites/alisoncoleman/2014/02/14/e
    ntrepreneur-the-french-do-have-a-word-for-it/>.

"Cost of Living Comparison Between San Francisco, CA and
    Sydney." *Numbeo*. www.numbeo.com, *n.d*. Web. 22 Oct.
    2014. <http://www.numbeo.com/cost-of-
    living/compare_cities.jsp?country1=United
    States&country2=Australia&city1=San Francisco,
    CA&city2=Sydney>.

Davis, Kathleen. "The World's 20 Hottest Startup Scenes
    (Infographic)."*Entrepreneur*. www.entrepreneur.com, 14
    Aug. 2013. Web. 24 Oct. 2014.
    <http://www.entrepreneur.com/article/227832>.

Desjardins, Jeff. "Infographic: Is Vancouver a Legitimate Tech
    Hub?" *Visual Capitalist*. www.visualcapitalist.com, 3 July

2014. Web. 22 Oct. 2014.

<http://www.visualcapitalist.com/infographic-vancouver-

legitimate-tech-hub/>.

Dillet, Romain. "Welcome To The French Tech

Ecosystem." *TechCruch*. www.techcrunch.com, 29 Jan. 2014.

Web. 24 Oct. 2014.

<http://techcrunch.com/2014/01/29/welcome-to-the-

french-tech-ecosystem/>.

Empson, Rip. "Startup Genome Ranks The World's Top Startup

Ecosystems: Silicon Valley, Tel Aviv & L.A. Lead The

Way."*TechCrunch*. www.techcrunch.com, 20 Nov. 2012.

Web. 24 Oct. 2014.

<http://techcrunch.com/2012/11/20/startup-genome-ranks-

the-worlds-top-startup-ecosystems-silicon-valley-tel-aviv-l-a-

lead-the-way/>.

Fannin, Rebecca. "Sydney Could Emerge As An Equal to San

Francisco's Tech Scene If Only..." *Forbes*. Forbes Magazine, 5

May 2014. Web. 22 Oct. 2014.

<http://www.forbes.com/sites/rebeccafannin/2014/05/05/s

ydney-could-emerge-as-an-equal-to-san-francisos-tech-

scene-if-only/>.

Fehrenbacher, Katie. "10 Things to Know about Tech Startups in

Brazil."*Gigaom*. www.gigeom.com, 17 May 2012. Web. 24

Oct. 2014. <https://gigaom.com/2012/05/17/10-things-to-

know-about-tech-startups-in-brazil/>.

Fehrenbacher, Katie. "London's Tech Startup Scene Is Hot —

Just Don't Compare It to Silicon Valley." *Gigaom*.

www.gigeom.com, 21 June 2013. Web. 24 Oct. 2014.

<http://gigaom.com/2013/06/21/londons-tech-startup-

scene-is-hot-just-dont-compare-it-to-silicon-valley/>.

The Firehose Project. www.thefirehoseproject.com, *n.d*. Web.

22 Oct. 2014. <http://www.thefirehoseproject.com/>.

Fishbein, Rebecca. "Is San Francisco A More Expensive City Than

   NYC?"*Gothamist*. http://gothamist.com/, 25 Aug. 2012.

   Web. 24 Oct. 2014.

   <http://gothamist.com/2012/08/25/will_san_francisco_beat

   _new_york_as.php>.

Fong, Cherise. "Brazil's High-tech Hub Grows in Sao Paulo's

   Brooklin."*CNN*. Cable News Network, 9 Apr. 2009. Web. 24

   Oct. 2014.

   <http://edition.cnn.com/2009/TECH/04/08/digitalbiz.berrini

   />.

Gerard, Shane. "Toronto Named Canada's High Tech Hub."

   Toronto. wx.toronto.ca, 30 Mar. 2011. Web. 24 Oct. 2014.

   <http://wx.toronto.ca/inter/it/newsrel.nsf/bydate/B6C13EB

   6848FD0F685257863006BA3AB>.

Geromel, Ricardo. "Top 10 Startups in Brazil." *Forbes*. Forbes

   Magazine, 20 Oct. 2011. Web. 24 Oct. 2014.

<http://www.forbes.com/sites/ricardogeromel/2011/10/20/
top-10-startups-in-brazil/>.

Gray, Bridget. *Going Global? Exploring Sydney Australia*.
bridgingthepacific.us, 11 Feb. 2014. Web. 22 Oct. 2014.
<http://bridgingthepacific.us/2014/02/11/tips-for-us-
companies-hiring-the-best-tech-talent-in-australia/>.

Hannon, Jenna. "Keep It Down Up There | The Canadian Tech
Scene in Toronto." *Techzulu*. http://techzulu.com, 17 Apr.
2012. Web. 24 Oct. 2014. <http://techzulu.com/keep-it-
down-up-there-the-canadian-tech-scene-in-toronto/>.

Heim, Anna. "18 Latin American Tech Hubs You Should
Know." *TNW Network All Stories RSS*. www.thenextweb.com,
11 Aug. 2012. Web. 24 Oct. 2014.

Heim, Anna. "Inside 9 Awesome Tech Workplaces in Sao
Paulo." *TNW Network All Stories RSS*. www.thenextweb.com,
28 Feb. 2013. Web. 24 Oct. 2014.

<http://thenextweb.com/la/2013/02/28/coolest-offices-

inside-9-awesome-tech-workplaces-in-sao-paulo/1/>.

Islam, Shafqat. "Forget Silicon Valley: Tech CEO Explains What's

So Great About New York." *Betabeat*. www.betabeat.com,

n.d. Web. 24 Oct. 2014.

<http://betabeat.com/2014/06/forget-silicon-valley-tech-

ceo-explains-whats-so-great-about-new-york/>.

Johnson, Bobbie. "London's Tech Scene Is Having Its Golden

Moment."*Bloomburg Businessweek Technology*.

http://www.businessweek.com, 27 July 2012. Web. 24 Oct.

2014. <http://www.businessweek.com/articles/2012-07-

27/londons-tech-scene-is-having-its-golden-moment>.

Keane, Bryce. "Starting Up Down Under: The Guide to

Australia's Tech Scene." *TNW Network All Stories RSS*.

thenextweb.com, 16 Jan. 2014. Web. 22 Oct. 2014.

<http://thenextweb.com/au/2014/01/16/starting-ultimate-

guide-australias-growing-startup-scene/>.

Kirsner, Scott. "Field Guide to Boston's Tech Scene - The Boston

    Globe."*The Boston Globe*. www.bostonglobe.com, 25 May

    2014. Web. 24 Oct. 2014.

    <http://www.bostonglobe.com/business/2014/05/24/field-

    guide-boston-tech-

    scene/MySSa3ao8gjPe7f65fvjAK/story.html>.

Le Wagon. www.lewagon.org, *n.d.* Web. 22 Oct. 2014.

    <http://www.lewagon.org/>.

Les Incubateurs Parisiens. Paris.

Lindzon, Jared. "THREE CANADIAN CITIES RANKED TOP 20 MOST

    ACTIVE STARTUP SCENES (INFOGRAPHIC)." *Betakit*.

    www.betakit.com, 16 Aug. 2013. Web. 24 Oct. 2014.

    <http://www.betakit.com/three-canadian-cities-ranked-top-

    20-most-active-startup-scenes-infographic/>.

"The London Startup Scene: Too Much Funding, Boozing and

    Not Enough Collaboration and Execution." *TechCrunch*.

http://techcrunch.com, 23 Sept. 2010. Web. 24 Oct. 2014.

<http://techcrunch.com/2010/09/23/the-london-startup-

scene-too-much-funding-boozing-and-not-enough-

collaboration-and-execution/>.

Mccarthy, Paul. "What Is the Tech Hub / Silicon Valley of

Australia?" *Quora*. www.quora.com, *n.d*. Web. 22 Oct. 2014.

<http://www.quora.com/What-is-the-Tech-hub-Silicon-

Valley-of-Australia>.

Medeiros, Joao. "Europe's Hottest Startup Capitals: Paris (Wired

UK)."*Wired UK*. www.wired.co.uk, 15 Aug. 2011. Web. 24

Oct. 2014.

<http://www.wired.co.uk/magazine/archive/2011/09/europ

ean-startups/paris>.

Merino, Faith. "Top Accelerators and Incubators in

London." *VatorNews*. http://vator.tv, 12 July 2013. Web. 24

Oct. 2014. <http://vator.tv/news/2013-07-12-top-

accelerators-and-incubators-in-london>.

Munford, Monty. "With 5,000 Startups, Tel Aviv Is Edging Into

the Tech Spotlight." *Mashable*. www.mashable.com, 17 Sept.

2013. Web. 24 Oct. 2014.

<http://mashable.com/2013/09/17/tel-aviv-tech/>.

O'Neal, Patricia. "Boston: Reclaiming Its Place as a Technology

Hub | Xconomy." *Xconomy RSS*. http://www.xconomy.com,

26 Apr. 2013. Web. 24 Oct. 2014.

<http://www.xconomy.com/boston/2013/04/26/boston-

reclaiming-its-place-as-a-technology-hub/>.

Peng, Wilson. "How Much Does It Cost To Live In The Silicon

Valley?" *Entrepreneur Sky: Startup & Tech Buzz*.

http://entrepreneursky.com, 18 Mar. 2014. Web. 24 Oct.

2014. <http://entrepreneursky.com/much-cost-live-silicon-

valley/>.

"PM Announces £50m Funding to Regenerate Old Street

Roundabout."*Gov.uk*. https://www.gov.uk, 6 Dec. 2012.

Web. 24 Oct. 2014.

<https://www.gov.uk/government/news/pm-announces-

50m-funding-to-regenerate-old-street-roundabout>.

Rapacon, Stacy. "Most Expensive U.S. Cities to Live

In." *Kiplinger*. www.kiplinger.com, 1 May 2014. Web. 24 Oct.

2014. <http://www.kiplinger.com/slideshow/real-

estate/T006-S001-most-expensive-u-s-cities-to-live-in/>.

Roughol, Isabelle. "LeWeb: A Traveler's Guide to Europe's Tech

Scene."*Linkedin*. www.linkedin.com, 10 Dec. 2013. Web. 24

Oct. 2014.

<https://www.linkedin.com/today/post/article/2013121006

2043-13333827-a-traveler-s-guide-to-europe-s-tech-scene>.

Salkever, Alex, and Vivek Wadhwa. "Why Entrepreneurship

Needs Immigrants." *Inc.* www.inc.com, 15 Oct. 2012. Web.

24 Oct. 2014. <http://www.inc.com/alex-salkever/why-

entrepreneurship-needs-immigrants.html>.

Sharma, Mahesh. "Silicon Hills? It Has a Ring to It." *The Sydney Morning Herald: Itpro*. http://www.smh.com.au, 15 June 2012. Web. 22 Oct. 2014. <http://www.smh.com.au/it-pro/business-it/silicon-hills-it-has-a-ring-to-it-20120615-20egj.html>.

"Startup Ecosystem Report 2012." *Multisite -Blog*. Telefonica Digital, *n.d*. Web. 22 Oct. 2014. <http://multisite-blog.digital.telefonica.com.s3.amazonaws.com/wp-content/uploads/2012/11/StartupEcosystemReport2012-PressSummary.pdf>.

"Startup Ecosystem Report 2012: Part One." *Multisite -Blog*. Telefonica Digital, *n.d*. Web. 22 Oct. 2014. <http://multisite-blog.digital.telefonica.com.s3.amazonaws.com/wp-content/uploads/2013/01/Startup-Eco_14012013.pdf>.

Strauss, Karsten. "The World's Top 4 Tech Capitals To Watch (after Silicon Valley and New York)." *Forbes*. Forbes Magazine, 20 Mar. 2013. Web. 24 Oct. 2014.

<http://www.forbes.com/sites/karstenstrauss/2013/03/20/t

he-worlds-top-4-tech-capitals-to-watch-after-silicon-valley-

and-new-york/2/>.

Sullivan, Justin. "4. San Francisco." *CBSNews*. CBS Interactive,

n.d. Web. 24 Oct. 2014.

<http://www.cbsnews.com/pictures/10-most-expensive-

cities-in-america/5/>.

Taylor, Colleen. "A Trip To MaRS, Downtown Toronto's Massive

Tech Hub [TCTV]." *TechCrunch*. http://techcrunch.com/, 16

Nov. 2012. Web. 24 Oct. 2014.

<http://techcrunch.com/2012/11/16/a-trip-to-mars-

downtown-torontos-massive-tech-hub-tctv/>.

Technology News Vancouver. www.techvibes.com, *n.d*. Web. 22

Oct. 2014. <http://www.techvibes.com/vancouver>.

Toscano, Nick. "Minimum Wage up 3 per Cent, Rise of $18.70 a

Week." *The Sydney Morning Herald*. www.smh.com, 4 June

2014. Web. 22 Oct. 2014.

<http://www.smh.com.au/national/minimum-wage-up-3-

per-cent-rise-of-1870-a-week-20140604-39is5.html>.

Waddell, Nick. "Vancouver's Tech Scene Emerges from the

Drizzle." *Cantech Letter*. www.cantechletter.com, 13 Feb.

2013. Web. 22 Oct. 2014.

<http://www.cantechletter.com/2013/02/vancouvers-tech-

scene-emerges-from-the-drizzle/>.

"Web Developer Salary." *Payscale. Human.

Capital.* www.payscale.com, *n.d.* Web. 22 Oct. 2014.

<http://www.payscale.com/research/CA/Job=Web_Develop

er/Salary>.

What the Job Market Wants. New York.

Zeveloff, Julie. "The 10 Most Expensive Cities In The United

States." *Business Insider*. Business Insider, Inc, 12 Feb. 2013.

Web. 24 Oct. 2014. <http://www.businessinsider.com/most-expensive-urban-areas-in-america-2013-2?op=1>.

What the Job Market Wants. 2012. The New York Times, New York. Web. 27 October 2014.

MBA VS...

"Educational Attainment in the United States: 2013." *United States Census Bureau*. United States Government, *n.d*. Web. 25 Oct. 2014. <https://www.census.gov/hhes/socdemo/education/data/cps/2013/tables.html>.

Eggleston, Liz. "Course Report Bootcamp Graduate Demographics & Outcomes Study." *Course Report*. Course Guide Inc., *n.d*. Web. 22 Oct. 2014. <https://www.coursereport.com/resources/course-report-bootcamp-graduate-demographics-outcomes-study>.

"Employment Statistics." *Yale School Of Management*. Yale, *n.d.*

   Web. 25 Oct. 2014. <http://som.yale.edu/our-

   programs/mba/careers/employment-statistics>.

Jacobs, Peter. "The College Majors With The Biggest

   Lifetime Earnings."*Business Insider*. Business

   Insider, Inc, 29 Sept. 2014. Web. 01 Nov. 2014.

Julien, Tiffany. "Work-Life Earnings by Field of Degree and

   Occupation for People With a Bachelor's Degree:

   2011." *Census*. United States Government, 1 Oct. 2012. Web.

   25 Oct. 2014.

   <http://www.census.gov/prod/2012pubs/acsbr11-04.pdf>.

O'Conner, Shawn. "Grad School: Still Worth the

   Money?" *Forbes*. Forbes Magazine, 5 Apr. 2012. Web. 25

   Oct. 2014.

   <http://www.forbes.com/sites/shawnoconnor/2012/04/05/

   grad-school-still-worth-the-money/>.

Stephens, Dale. "A Smart Investor Would Skip the M.B.A." *The Wall Street Journal Life & Culture*. Dow Jones & Company, 1 Mar. 2013. Web. 25 Oct. 2014. <http://online.wsj.com/news/articles/SB10001424127887323884304578328243334068564>.

Time, Forest. "The Starting Salary for Web Development." *Chron*. Houston Chronicle, *n.d*. Web. 25 Oct. 2014. <http://work.chron.com/starting-salary-development-10848.html>.

Weiss, Tara. "M.B.A. Reality Check." *Forbes*. Forbes Magazine, 1 Aug. 2006. Web. 25 Oct. 2014. <http://www.forbes.com/2006/08/01/leadership-mba-salary-cx_tw_0801mbacomp.html>.

Zlomek, Erin. "More MBA Grads Are Switching Careers as Job Market Improves." *Bloomberg Business Week*. Bloomberg, 18 Apr. 2013. Web. 25 Oct. 2014.

<http://www.businessweek.com/articles/2013-04-18/more-

mba-grads-are-switching-careers-as-job-market-improves>.

A Classic C.S. Degree...

"The Unofficial Guide to Computer Science @ Harvard." *Harvard

School of Engineering and Applied Sciences*. Harvard, *n.d.*

Web. 25 Oct. 2014. <http://guide.cs50.net/guide-8.pdf>.

Your First Programs...

"Quiz FAQ." *Ruby Quiz*. http://rubyquiz.com, *n.d.*. Web. 25 Oct.

2014. <http://rubyquiz.com/>.

Demand for Programmers

Andreessen, Marc. "Why Software Is Eating The World." *The

Wall Street Journal Life & Culture*. Dow Jones & Company, 20

Aug. 2011. Web. 25 Oct. 2014.

<http://online.wsj.com/news/articles/SB1000142405311190
348090457651225091562946O>.

Ballenstedt, Brittany. "IT Jobless Rates Are Half the National
Average." *Wired Workplace*. Nextgov, 6 July 2012. Web. 25
Oct. 2014. <http://www.nextgov.com/cio-briefing/wired-
workplace/2012/07/it-jobless-rates-are-half-national-
average/56656/>.

Bersin, Josh. "The Software Economy: Why Software Jobs Are
Taking Over." *Forbes*. Forbes Magazine, 5 Aug. 2013. Web.
25 Oct. 2014.
<http://www.forbes.com/sites/joshbersin/2013/08/05/the-
software-economy-why-software-jobs-are-taking-over/>.

"The Best Jobs for 2014." *Reboot Illinois*.
http://www.rebootillinois.com, *n.d*. Web. 25 Oct. 2014.
<http://www.rebootillinois.com/Opinion/the-best-jobs-for-
2014-11679>.

"Best Technology Jobs." *Business News and Financial News*. US

News, *n.d.* Web. 25 Oct. 2014.

<http://money.usnews.com/careers/best-

jobs/rankings/best-technology-jobs>.

"Computer Programmers." *U.S. Bureau of Labor Statistics*. U.S.

Bureau of Labor Statistics, 8 Jan. 2014. Web. 25 Oct. 2014.

<http://www.bls.gov/ooh/computer-and-information-

technology/computer-programmers.htm#tab-6>.

Craig, Caroline. "Silicon Valley: Top Salaries for Many, Greater

Inequality for All." *InfoWorld*. www.infoworld.com, 18 Oct.

2013. Web. 25 Oct. 2014. <http://www.infoworld.com/t/it-

jobs/silicon-valley-top-salaries-many-greater-inequality-all-

229000>.

"Demographics." *U.S. Bureau of Labor Statistics*. U.S. Bureau of

Labor Statistics, *n.d.* Web. 25 Oct. 2014.

<http://www.bls.gov/cps/demographics.htm>.

Dietrich, Matthew. "The Best Jobs for 2014." *The Huffington Post*. TheHuffingtonPost.com, 22 Jan. 2014. Web. 25 Oct. 2014. <http://www.huffingtonpost.com/matthew-dietrich/the-best-jobs-for-2014_b_4645511.html>.

Evans, Jon. "How Long Will Programmers Be So Well-Paid?" *TechCrunch*. www.techcrunch.com, 27 Oct. 2012. Web. 25 Oct. 2014. <http://techcrunch.com/2012/10/27/write-code-get-paid/>.

Hargreaves, Steve. "Jobs with the Lowest (and Highest) Unemployment Rates." *CNNMoney*. Cable News Network, 4 Jan. 2013. Web. 25 Oct. 2014. <http://money.cnn.com/2013/01/04/news/economy/jobs-lowest-unemployment/>.

Hellman, Nathan. "Best Technology Jobs Software Developer." *US News Money*. US News, *n.d.*. Web. 25 Oct. 2014. <http://money.usnews.com/careers/best-jobs/software-developer>.

"Infographic: Just How Big Is the Nerd Economy?" *New Relic Blog*. http://blog.newrelic.com, 25 Oct. 2012. Web. 25 Oct. 2014. <http://blog.newrelic.com/2012/10/25/infographic-just-how-big-is-the-nerd-economy/>.

Jolly, David. "Unemployment in Europe Stays High Amid Signs of Recovery." *The New York Times*. The New York Times, 8 Jan. 2014. Web. 25 Oct. 2014. <http://www.nytimes.com/2014/01/09/business/internatio nal/unemployment-in-euro-zone.html?_r=0>.

Kelly, Samantha. "Top Tech Jobs in Demand and Their Salaries [INFOGRAPHIC]." *Mashable*. www.mashable.com, 4 Oct. 2012. Web. 25 Oct. 2014. <http://mashable.com/2012/10/04/tech-jobs-in-demand/>.

Knorr, Eric. "Now the Developer Is King." *InfoWorld*. www.infoworld.com, 1 July 2013. Web. 25 Oct. 2014. <http://www.infoworld.com/t/application-development/now-the-developer-king-221850>.

Meyer, Ann. "Software Engineers Hard to Find." *Chicago
Tribune*. www.chicagotribune.com, 17 Oct. 2010. Web. 25
Oct. 2014. <http://articles.chicagotribune.com/2010-10-
17/business/ct-biz-1017-out-technology-
20101017_1_software-engineers-chicago-developers-tech-
jobs/2>.

"Money." *MSN*. www.msn.com, *n.d*. Web. 25 Oct. 2014.
<http://money.msn.com/personal-finance/the-10-best-jobs-
of-2014-1>.

Samson, Ted. "IT Jobs Light up Top 100 Careers for
2013." *InfoWorld*. www.infoworld.com, 20 Dec. 2012. Web.
25 Oct. 2014. <http://www.infoworld.com/t/it-jobs/it-jobs-
light-top-100-careers-2013-209585>.

Smith, Jacquelyn. "The Top Jobs For 2014." *Forbes*. Forbes
Magazine, 12 Dec. 2013. Web. 25 Oct. 2014.
<http://www.forbes.com/sites/jacquelynsmith/2013/12/12/
the-top-jobs-for-2014/>.

"Software Developers." *U.S. Bureau of Labor Statistics*. U.S.

    Bureau of Labor Statistics, 8 Jan. 2014. Web. 25 Oct. 2014.

    <http://www.bls.gov/ooh/computer-and-information-

    technology/software-developers.htm#tab-1>.

"Summary Report for Computer Programmers." *O*Net Online*.

    http://www.onetonline.org, 1 Jan. 2010. Web. 25 Oct. 2014.

    <http://www.onetonline.org/link/summary/15-1131.00>.

"Summary Report for Software Developers, Systems

    Software." *O*Net Online*. http://www.onetonline.org, 1 Jan.

    2010. Web. 25 Oct. 2014.

    <http://www.onetonline.org/link/summary/15-1133.00>.

"Summary Report for Web Developers." *O*Net Online*.

    http://www.onetonline.org, 1 Jan. 2010. Web. 25 Oct. 2014.

    <http://www.onetonline.org/link/summary/15-1134.00>.

Tynan, Dan. "Tech Boom! The War for Top Developer

    Talent." *Javaworld*. www.javaworld.com, 2 Dec. 2013. Web.

25 Oct. 2014.

<http://www.javaworld.com/article/2078925/mobile-java/tech-boom--the-war-for-top-developer-talent.html>.

"In the Year 2016: The 30 Fastest-growing Jobs." *Boston.com*. The New York Times, 25 Dec. 2013. Web. 25 Oct. 2014. <http://www.boston.com/jobs/2013/12/23/the-year-the-fastest-growing-jobs/GGDo2PVgihmpsurdiPGUhO/story.html?pg=31>.

My Experience...

Davis, Kathleen. "The World's 20 Hottest Startup Scenes (Infographic)."*Entrepreneur*. www.entrepreneur.com, 14 Aug. 2013. Web. 25 Oct. 2014. <http://www.entrepreneur.com/article/227832>.

Dumon, Marv. "The Real Cost Of An MBA." *Investopedia*. www.investopedia.com, *n.d*. Web. 25 Oct. 2014.

<http://www.investopedia.com/articles/professionaleducati
on/09/mba-real-costs.asp>.

Graham, Jefferson. "Top Cities for Technology Start-ups." *USA
Today Tech*. USA Today, 24 Aug. 2012. Web. 25 Oct. 2014.
<http://usatoday30.usatoday.com/tech/columnist/talkingtec
h/story/2012-08-22/top-tech-startup-cities/57220670/1>.

"The MBA Myth." *Signal vs. Noise by Basecamp — Business,
Design, Programming and the Web*.
https://signalvnoise.com, 26 June 2008. Web. 25 Oct. 2014.
<https://signalvnoise.com/posts/1112-the-mba-myth>.

Meng, Victoria. "The Ultimate Guide to Coding Bootcamps: The
Most Selective Bootcamps." *Skilledup*. www.skilledup.com,
n.d. Web. 25 Oct. 2014.
<http://www.skilledup.com/learn/programming/the-
ultimate-guide-to-coding-bootcamps-the-most-selective-
bootcamps/>.

"Now for Hire Full-Stack Web Developers." *Hire a Web Developer*. Launch Academy, *n.d.*. Web. 25 Oct. 2014. <http://www.launchacademy.com/hire_developer>.

O'Conner, Shawn. "Grad School: Still Worth the Money?" *Forbes*. Forbes Magazine, 5 Apr. 2012. Web. 25 Oct. 2014. <http://www.forbes.com/sites/shawnoconnor/2012/04/05/grad-school-still-worth-the-money/>.

Schweitzer, Karen. "What Is the Average Cost of an MBA Degree?" *About Education*. About, *n.d.*. Web. 25 Oct. 2014. <http://businessmajors.about.com/od/bschoolrankings/a/How-Much-Does-It-Cost-To-Earn-An-Mba-Degree.htm>.

Stephens, Dale. "A Smart Investor Would Skip the M.B.A." *The Wall Street Journal*. Dow Jones & Company, 1 Mar. 2013. Web. 25 Oct. 2014. <http://online.wsj.com/news/articles/SB1000142412788732388430457832824334068564>.

"Table A-12. Unemployed Persons by Duration of

Unemployment." *U.S. Bureau of Labor Statistics*. U.S. Bureau

of Labor Statistics, 3 Oct. 2014. Web. 25 Oct. 2014.

<http://www.bls.gov/news.release/empsit.t12.htm>.